Mass Customization

A Supply Chain Approach

Mass Customization

A Supply Chain Approach

Charu Chandra
University of Michigan–Dearborn
Dearborn, Michigan

and

Ali Kamrani
University of Houston
Houston, Texas

Kluwer Academic / Plenum Publishers
New York, Boston, Dordrecht, London, Moscow

Library of Congress Cataloging-in-Publication Data

Chandra, Charu.
    Mass customization: a supply chain approach/Charu Chandra, Ali Kamrani.
      p.   cm.
    Includes bibliographical references and index.
    ISBN 0-306-48505-2 (hardbound)
      1. Flexible manufacturing systems.   2. Product management.   3. Customer relations.   I.
Kamrani, Ali.   II. Title.

TS155.65.C48 2004
670.42′7—dc22

2004047320

ISBN 0-306-48505-2

©2004 Kluwer Academic/Plenum Publishers, New York
233 Spring Street, New York, New York 10013

http://www.kluweronline.com

10 9 8 7 6 5 4 3 2 1

A C.I.P. record for this book is available from the Library of Congress

Permissions for books published in Europe: *permissions@wkap.nl*
Permissions for books published in the United States of America: *permissions@wkap.com*

Printed in the United States of America

PREFACE

As firms seek to expand their market base in an increasingly consumer-focused environment, interest in mass customization as a manufacturing strategy, continues to grow, in industry and academia. A number of factors have contributed to this trend.

Consumers are demanding products that feature latest in style and technology; offer utility, value and price; and meet quality and reliability expectations. In order to meet consumer needs and yet stay competitive, firms have had to cluster products based on common product features and then map these on to identical processes and operations.

Managing effects on manufacturing productivity in the presence of these multi objectives, i.e., the impact of product variety management on time-to-market (lead-time, set up time), cost, and scaling of manufacturing / production operations (lot sizing), is a topic of much interest in supply chain management research. This edited book addresses some of these topics by offering a discussion on issues, problems, and solutions in managing a supply chain in a manufacturing enterprise, utilizing mass customization strategies.

The contributed chapters in this book offer a summary of the research in focus areas of mass customization; applicable frameworks, models, methodologies, and technologies; and prototypical applications to illustrate this complex, yet emerging field of inquiry. Contributions are organized in *four* main categories:

1. Concepts and state-of-research in mass customization.
2. Problem solving frameworks, models, and methodologies.
3. Supportive techniques and technologies for enabling mass customization.
4. Future research agenda.

We elaborate on each below.

1. Concepts and state-of-research in mass customization

Chapters in this category introduce basic concepts in mass customization, supply chain, and integrated logistics in supply chain utilizing mass customization strategy. Research in mass customization is categorized as an aid to problem solving.

In Chapter 1, Charu Chandra and Jānis Grabis introduce the mass customization strategy by comparing and contrasting it with traditional manufacturing. They investigate implications of this strategy on the entire operations of a manufacturing enterprise by presenting an overview of mass customization enablers, success factors, strategies, and approaches. This leads to a discussion on the influence of mass customization strategy, particularly on managing logistics in a supply chain.

In Chapter 2 by Charu Chandra and Jānis Grabis, categorization of mass customization research is compiled in order to provide an overall comprehension about research and application progress. Research articles have been categorized using mass customization strategy, enabler, analysis method and application area criteria.

2. Problem solving framework, models, and methodologies

Chapters in this category describe a problem solving framework for managing mass customizable manufacturing systems utilizing integrated supply chain approaches. Models for logistical planning both at macro and micro levels are described.

Chapter 3 by Charu Chandra and Jānis Grabis places mass customization in the wider perspective of visionary evolution of manufacturing systems. Problem-solving approaches for managing reconfigurable supply chains are discussed. A decision support system integrating the information support system and the decision modeling system is proposed.

Chapter 4 by Charu Chandra and Jānis Grabis emphasizes efficient logistics and supply chain management as one of the key preconditions for adopting mass customization strategies. Existing literature to analyze interrelationships between supply chain management and mass customization strategies is discussed. Models for supply chain configuration and inventory management addressing some of the key issues in logistics management for mass customization is discussed.

3. Supportive techniques and technologies for enabling mass customization

Chapters in this category describe supportive techniques and technologies for enabling mass customization in a manufacturing enterprise. A broad spectrum of advanced techniques for modularity, collaborative engineering, integrated product-process design, and production planning are presented. Use of information, simulation, and CAD modeling technologies are presented to apply these techniques.

In Chapter 5, Ali Kamrani and Sa'ed Salhieh describe the application of modularity in optimum design of a product or system under constraints of multi-objectives as imposed by the mass customization strategy.

In Chapter 6, Ali Kamrani deals with the issue of enhanced product variety and shortened time-to-market by proposing a methodology that emphasizes integration of software tools and resources involved in the design process to collaborate between geographically dispersed design teams and vendors.

In Chapter 7, Mary Meixell and David Wu examine the role of supply chain collaboration in a manufacturing environment where products are mass customized. Issues of supply chain coordination are investigated in applying postponement through product and process design.

In Chapter 8, Manfredi Bruccoleri describes the use of simulation in conjunction with multi-agent system techniques to develop realistic prototypes of complex decision support systems for the design of coordination strategies in large-scale mass customization production systems.

In Chapter 9, Charu Chandra and Ali Kamrani propose a knowledge management approach based on consumer-focused product development philosophy (a direct offshoot of the mass customization strategy). It integrates capabilities for (a) intelligent information support, and (b) group decision-making, utilizing a common enterprise network model and knowledge interface through shared ontologies.

4. *Future research agenda in mass customization*

Chapter 10 in this category, by Janet Efstathiou and Ting Zhang offers a glimpse of issues for future research in mass customization topic. It highlights the need for developing scientific models of mass customization that integrate a manufacturing enterprise's core competencies, performance enablers, and available strategies.

This book is written to serve as a textbook for advanced undergraduate and beginning graduate studies in Operations Management, Management Science, and Industrial and Systems Engineering. It can also serve as a reference for researchers and practitioners.

Charu Chandra
Dearborn, Michigan

Ali K. Kamrani
Houston, Texas

ACKNOWLEDGMENTS

We gratefully acknowledge all those who helped us in bringing out this edited book for publication. First and foremost we thank contributors to various chapters, without whose efforts this book would not have been possible.

Our sincere thanks to Mr. Armen Tumanyan, Graduate Student Research Student at University of Michigan – Dearborn for his help in formatting the book.

Finally, we wish to thank our editor, Mr. Ray O'Connell and his assistant Ms. Corinne D'Italia for their encouragement and support in successful completion of this book.

CONTENTS

7. COLLABORATIVE MANUFACTURING FOR MASS CUSTOMIZATION 169

MARY J. MEIXELL, S. DAVID WU

8. SIMULATION MODELING USING AGENTS FOR MASS CUSTOMIZATION 191

MANFREDI BRUCCOLERI

9. KNOWLEDGE MANAGEMENT FOR CONSUMER-FOCUSED PRODUCT DESIGN 211

CHARU CHANDRA, ALI K. KAMRANI

SECTION 4: FUTURE RESEARCH AGENDA 235

10. FUTURE DIRECTION ON MASS CUSTOMIZATION RESEARCH AND PRACTICES: A RESEARCH AGENDA 237

JANET EFSTATHIOU, TING ZHANG

LIST OF FIGURES

LIST OF TABLES

CONTRIBUTORS

Manfredi Bruccoleri
Dipartimento di Tecnologia
  Meccanica
Produzione e Ingegneria Gestionale
Faculty of Engineering
University of Palermo
ITALY

Charu Chandra
Industrial and Manufacturing
Systems Engineering Department
University of Michigan – Dearborn
Dearborn, Michigan, U.S.A.

Janet Efstathiou
Manufacturing Systems Group
Department of Engineering Science
University of Oxford
Oxford, U.K.

Jānis Grabis
Department of Operations Research
Institute of Information Technology
Riga Technical University
Riga, Latvia

Ali K. Kamrani
Industrial Engineering Department
University of Houston
Houston, Texas, U.S.A.

Mahesh Kanawade
Manufacturing Systems Engineering
  Department
University of Michigan – Dearborn
Dearborn, Michigan, U.S.A.

Mary J. Meixell
Department of Decision Sciences
School of Management
George Mason University
Fairfax, Virginia, U.S.A.

Sa'ed M. Salhieh
Industrial Engineering Department
The University of Jordan
Amman, Jordan

S. David Wu
Industrial and Systems Engineering
Lehigh University
Bethlehem, Pennsylvania, U.S.A.

Ting Zhang
Manufacturing Systems Group
Department of Engineering Science
University of Oxford
Oxford, U.K.

SECTION 1:

CONCEPTS AND STATE-OF-RESEARCH IN MASS CUSTOMIZATION

CHAPTER 1

MANAGING LOGISTICS FOR MASS CUSTOMIZATION: THE NEW PRODUCTION FRONTIER

Charu Chandra[1], Jānis Grabis[2]

[1]*University of Michigan-Dearborn*
[2]*Riga Technical University*

Abstract: Combining aspects of traditional manufacturing techniques, the topic of mass customization is being pursued in research and industry. With consumers demanding specialized products, firms are employing mass customization strategy, which has implications on the entire operations of an enterprise. This Chapter investigates presents an overview of mass customization enablers, success factors, strategies, and approaches. Exploring these issues leads to discussion on the influence of mass customization strategy, particularly for managing logistics in a supply chain.

Keywords: Mass customization, logistics management, manufacturing strategy, product family.

1. INTRODUCTION

For a better part of last century, mass production was the key manufacturing strategy in efficiently producing products and services at very low cost. Mass production systems are characterized by their ability to produce large volumes of standardized products at low cost, achieved through repetitive operations and long running production lines. In such a system, production is pushed through long running lines according to sales forecasts and material inventory levels. However, with shortened life cycles and rapid change in today's market, mass production systems are not capable of responding quickly. In order to resolve this lack of responsiveness, many firms have made the invaluable shift from material requirements planning (MRP) to enterprise-wide requirements planning (ERP) which provides informative and timely demand data, utilized by companies to plan production runs that parallel actual demand.

With these new production capabilities, firms wish to be able to accommodate consumers' growing needs for product variety and flexibility in order to remain competitive. To attain the flexibility and responsiveness needed, many firms have adopted the emerging paradigm of mass customization. In general, mass customization systems enable the production of cost-effective customized products and services.

Many researchers and industry professionals are investigating mass customization topic because of its innovative capabilities and cross-discipline research focus. This Chapter provides an introductory overview of mass customization in order to understand its scope, complexities and practical importance. Section 2 offers an introduction to mass customization covering historical trends and opportunities, factors to consider for implementing mass customization strategy and enablers of mass customization. Section 3 offers a discussion on mass customization strategy. Section 4 highlights some of the key issues in implementing mass customization strategy.

2. MASS CUSTOMIZATION: A NEW PARADIGM IN MANUFACTURING

Two distinct concepts, the visionary and the practical concepts define mass customization. The broader, visionary concept defines mass customization as the ability to profitably provide customers with individually designed products and services anytime, anywhere and anyway they want (Da Silveira et al. 2001). The narrower, practical concept defines mass customization as the use of flexible processes and organizational structures to provide a variety of products and services that are designed to individual customer specifications, near the cost of a mass production system (Hart 1995).

In order to distinctly characterize a mass customization system in comparison with other manufacturing systems, its position on the Hayes and Wheelwright (1979) product-process matrix may be examined, as found in Table 1-1. The product-process matrix was developed to illustrate the interaction between product and process life cycle stages. It serves as a useful tool for showing how a firm's position on the matrix reflects its strengths and weaknesses. The vertical position on the matrix represents the process life cycle stages, while the horizontal position represents the product life cycle stages. Processes that are towards the top of the matrix are in fluid form, meaning they are more flexible but not very cost efficient. As one moves towards the bottom rows of the matrix, systematic and automated processes are represented. These processes are efficient and less costly than the more flexible processes. A firm can be characterized by the region it occupies on the matrix, determined by the stage of the product life cycle and the choice of production process selected for its products. Most manufacturing firms fall along the diagonal of the matrix, implying that they use traditional manufacturing techniques for production.

Table 1-1. Industry positions in product-process life cycle
(adopted from Hayes and Wheelwright (1979)

Product→ Process ↓	Low volume, custom products	Low-medium volume, many products	Medium-high volume, a few standardized products	Very high volume, a couple of commodity type products
Job shop	Aerospace			
Batch		Industrial equipment		
Assembly		Main mass customization opportunities	Automobiles, Appliances	
Continuous				Oil

As can be seen in the table, mass customization systems are positioned below the main diagonal of the matrix. These systems take the better of two starkly different options, having assembly line processes that are able to deliver higher product varieties.

2.1   Key factors for success of a mass customization system

The success of a mass customization system depends on several crucial factors. These factors discussed below can be analyzed with respect to four main categories: customer sensitivity, process amenability, competitive environment, and organizational readiness (Hart 1995).

Customer Sensitivity

Customer demand for individualized and customized products must exist for the success of a mass customization system and customers' demand for customized products is dependent upon two main factors. The first, degree of sacrifice the customer is willing to make for the customized product. It involves how much the customer is willing to pay and how long she/he is willing to wait. The second, the firm's ability to produce according to customer specifications, within a reasonable time and cost. The balance between these two factors determines the success of the mass customization system. The Motorola case involving portable communication devices (Eastwood 1996) demonstrates emergence of demand for customized product as customers get acquainted with the technology. In response, the firm reconfigures its operations to meet challenges of cost cutting and lead time reduction. However, the drive for mass customization needs to maintain its customer focus. Unjustified variety not only causes unnecessary costs, but also creates customer relation confusion (Huffman and Kahn 1998).

Process Amenability

Process amenability is very important to the development of a mass customization system. This category encompasses several areas including enablers, product design, and production design. Technological and organizational enablers include advanced manufacturing technology and information technologies that must be available for the implementation of a mass customization system. Enablers of mass customization are discussed in further detail in the subsequent section. With respect to the product and production design, products must be designed to be customizable. In other words, the firm must have access to information concerning individual needs and furthermore they must have processes in place to be able to translate these needs into actual specifications (Hart 1995). Consequently, much of mass customization research is devoted to capturing customer requirements and transformation of these requirements to product design manufacturability (Tseng and Jiao 1998, Aldanondo et al. 2003).

Competitive Environment

Market conditions must be appropriate to support a competitive environment. In this case, timing of the development of a mass customization system is a crucial factor. Market turbulence as defined by instability and unpredictability of demand, is very important in determining the right time to employ a mass customization system. The greater the market turbulence, the greater the potential customer need for variety and customization (Pine et al. 1993). Another important consideration of time is being the first to develop a mass customization system in any particular industry. It allows the firm to gain a competitive advantage, besides being innovative and customer-driven (Da Silveira et al. 2001). However, as mass customization is increasingly becoming more common, there are fewer opportunities for starting first. Competition is shifting towards greater focus on improved customization of most valuable features.

Organizational Readiness

Organizational readiness has to do with attitudes, culture and resources of the firm. Foremost, the firm's leadership capability is an important consideration. It must be enlightened, open to new ideas and aggressive in the pursuit of competitive advantage (Hart 1995), and promote a culture that emphasizes knowledge sharing through the development of networks and new product and process technologies. This type of culture is very important in developing the firm's ability to translate customer demands into products and services, especially since mass customization is a chain based concept (Feitzinger and Lee 1997). Its success greatly depends on the linked network and readiness of suppliers, manufacturers and retailers. Mass customization and technologies that are used to enable it require from a manufacturer, abilities to adapt to new requirements. Knowledge assumes a greater role than skills. The organizational structure is required to support knowledge generation and its efficient utilization (Hirschhorn et al. 2001). Additionally, mass customization also requires organizational changes at the shop-floor level (Gunasekaran et al. 1998).

2.2 Enablers of mass customization

Mass customization enablers are technologies and methodologies that support the development of the system (Da Silveira et al. 2001). Mass customization methodologies address organizational and cultural aspects of implementation, while process technologies address manufacturing aspects.

Agile manufacturing practices and supply chain management are two of the methodologies that serve as important tools for enabling mass customization. Agile manufacturing practices are characterized by their ability to prosper in rapidly changing environments, where the change is driven by customer demands for new products and product features. This requires re-programmable and re-configurable production systems that are able to economically operate with very small lot sizes (Da Silveira et al. 2001). There are certain interactions between agility and another mass customization enabler – lean manufacturing (Duguay et al. 1997). The lean manufacturing strategy, which is based on fine-tuning of every operation, works well if demand for customized products is stable. Agility is more appropriate for responding to highly uncertain customer demand with quickly changing preferences related to customized features.

Supply chain management has been referred to as the glue binding together activities performed for mass customization success (Gooley 1998). It is responsible for interactions with customers, distribution of customized products, manufacturing, purchasing materials from suppliers and coordination of supply chain operations. Efficiency of supply chain management is one of the major determinants in achieving the dual mass customization objectives of providing low cost and shorter delivery times. Mass customization is also one of the drivers forcing supply chain members to start coordination of their activities already at the product design phase (Salvador et al. 2002b).

Advanced manufacturing technologies such as flexible manufacturing systems, and network technologies such as computer-aided design and computer-integrated manufacturing are important tools for manufacturers moving toward mass customization. For example, flexible-manufacturing systems can manufacture assorted products with the same group of machines that are linked by automated material handling systems (Lau 1995). Communication and network technologies enable direct links between work groups for improvement of response time to customer requirements. Internet based information systems are a particular area of interest (Helander and Jiao 2002, Turowski 2002). These systems serve two primary objectives: 1) capturing customer requirements, and 2) coordination in the supply chain environment.

2.3 Industry Applications

Many companies and industries are adopting mass customization strategies to better meet customer requirements by offering products and services that were previously available only through a mass production environment. There are many examples of industries that have made this shift:

- Jeans – The Levi's Company offers "cut-to-fit" jeans that can be delivered to the customer's store. Customers can order their jeans according to their specifications including selected fabric, style, and personal dimensions. These custom fit jeans are made possible through flexible manufacturing processes. Unique patterns are built upon a traditional style pair of jeans that is altered to specific customer dimensions (Duray et al. 2000).

- Cellular phones and pagers – The Motorola Company has shifted from standardized communication devices to highly customized ones. Motorola allows customers to choose from a variety of features including language options, colors, and accessories to meet their specific needs. Motorola uses sophisticated information system to help customer select features from options that cover many combinations (Duray and Glen 1999).
- Computers – Dell Computer Company provides customization of personal computers through interchangeability of parts. Computer systems are assembled according to customer requirements by adding or subtracting components from one of the several base systems (Duray and Glen 1999). Different configurations are achieved through the combination of several parts including hard drives, chips, storage media, and accessories. These features are added during the assembly process.
- Passenger vehicles – Automotive companies excel in high volume manufacturing. However, customization is gradually gaining ground in this industry (Fisher and Ittner 1999, Griffiths and Margetts 2000). Customization and flexibility to respond to customer requirements need to be balanced with lean manufacturing practices. Fisher and Ittner (1999) describe assembly of customized vehicles according to the production schedule, indicating which features are needed for a particular vehicle. Customization is strictly limited to core components, such as body style and engine, while a large number of customizations can be assigned for peripherals. Core components are installed at the first two stages of the assembly. Peripherals are mainly installed during the final assembly stage. This is in line with classification of mass customization strategy in the automotive industry by Alford et al. 2000. They contend that standardized customization is most appropriate compared to pure customization (only for low volume vehicles) and cosmetic or segmented customization (difficult to meet customer requirements).

While industries have made tremendous progress in initiating mass customization strategies, however, the two potential challenges that still remain, are trying to maintain the lowest possible costs and lead times.

2.4   A Historical Perspective on Trends affecting Mass Customization Philosophy

In order to understand the significance of changes taking place in various manufacturing related initiatives including mass customization, it would be prudent to review historical aspects of production and operations management activities (Poirier and Reiter 1996).

During the period from 1960 to 1975, corporations had vertical organization structures and optimization of activities was focused on functions. Relationships with vendors were win-lose interaction, many a times adversarial. Manufacturing systems were focused on Materials Requirements Planning (MRP).

In the timeframe from 1975 to 1990, corporations were still vertically aligned, but several were involved in process mapping and analysis to evaluate their operations. There was realization by organizations of the benefit of integration of functions such as, product design and manufacturing. Various quality initiatives, such as the Total Quality Management philosophies of Deming, Juran, and Crosby, and ISO Standards for quality measurement, were initiated by many organizations. The Malcolm Baldridge award and Shingo Prize for recognizing excellence in these and other quality initiatives were initiated. Manufacturing systems were focused on MRP II.

Starting in 1990, corporations all over the world have been experiencing increasing national and international competition. Strategic alliances among organizations have been growing. Organization structures are starting to align with processes. Manufacturing systems in organizations have been enhanced with information technology tools such as, Enterprise Resource Planning, Distribution Requirements Planning, Electronic Commerce, Product Data Management, Collaborative Engineering, etc. (Aberdeen Group 1996). Design for disassembly, synchronous manufacturing, and agile manufacturing are some of the new paradigms in manufacturing. There has been a growing appreciation in many firms of total cost focus for a product from its source to consumption, as opposed to extracting lowest price from immediate vendor(s) (Turbide 1997). There has also been an increased reliance on purchased materials and outsourcing with a simultaneous reduction in the number of suppliers and greater sharing of information between vendors and customers. A noticeable shift has taken place in the marketplace from mass production to customized products. This has resulted in the emphasis on greater organizational and process flexibility and coordination of processes across many sites. More and more organizations are promoting employee empowerment and the need for rules-based, real-time decision support systems to attain organizational and process flexibility as well as responding to competitive pressure to introduce new products more quickly, cheaply and of improved quality.

The underlying philosophy of managing supply chains has evolved to respond to these changing business trends. Supply chain management phenomenon has received attention of researchers and practitioners in various topics. In the earlier years, the emphasis was on materials planning utilizing materials requirements planning techniques, inventory logistics management with one warehouse multi-retailer distribution system, and push and pull operation techniques for production systems. In the last few years, however, there has been a renewed interest in designing and implementing integrated systems, such as enterprise resource planning, multi-echelon inventory, and synchronous-flow manufacturing, respectively. A number of factors have contributed to this shift. First, there has been a realization that better planning and management of complex interrelated systems, such as materials planning, inventory management, capacity planning, logistics, and production systems will lead to overall improvement in enterprise productivity. Second, advances in information and communication technologies complemented by sophisticated decision support systems enable designing, implementing and controlling strategic and tactical strategies essential to delivery of integrated systems. The availability of such systems has the potential of fundamentally influencing enterprise integration issues. The motivation in describing research on mass customization issues related to supply chain is to propose a *framework* that enables dealing with such issues effectively.

The approach utilizing supply chain philosophy for enterprise integration described in this book, proposes domain independent problem solving and modeling, and domain dependent analysis and implementation. The purpose of the approach is to ascertain characteristics of the problem independent of the specific problem environment. Consequently, the approach delivers solution(s) or the solution method that are intrinsic to the problem and not its environment. Analysis methods help to understand characteristics of the solution methodology as well as provide specific guarantees of effectiveness. Invariably, insights gained from these analyses can be used to develop effective problem solving tools and techniques for complex enterprise integration problems that companies utilizing mass customization strategies encountered in the decision-making process.

2.5   Challenges and opportunities

Challenges and opportunities are viewed from the mass customization maturation perspective. Mass customization has become an established manufacturing strategy. It has overcome the initial promise of being exclusive alternative to mass production. It has been recognized that mass customization should be adopted only in response to real customer demand for customized products. Therefore, optimization of product variety and involvement of customers in product design will be gaining increasing attention. For instance, Fujita (2002) analyzes trade-off between variety and production costs, where variety increases potential sales and production costs. The involvement of customers in the product design allows better understanding of needed customized features. The challenge is avoiding increasing product development costs due to customer involvement by means of seamless shift of "development costs" to the customer side, especially, in the consumer market.

Maturation of mass customization also opens the door to finding synergies between mass customization and mass production. It has been found that firms' benefit from simultaneous production of both customized, and standardized products (Duray 2002). However, few investigations have addressed the problem of combining production of customized, and standardized products in a cost efficient manner. Similarly, postponement can be perceived as shift from the initial mass production part to the closing mass customization part. However, little is known about coordination between these two parts. For instance, the mass production part uses large batches for transporting sub-assemblies from suppliers to a manufacturer. Any delays can cause major disruption in the mass customization part. This problem relates to the final major challenge, viz., supply chain integration. Although, supply chain management is often mentioned as one of the most important mass customization enablers, supply chain structures considered are simple. Implementation of mass customization has led to consolidation of supply chain operations (Eastwood 1996, Kotha 1996) because of coordination and communication strains. This consolidation threatens with insufficient agility to respond to changing customer requirements and increasing competition. Rapid advances of both logical and infrastructural capabilities available for coordinated supply chain management (Min and Zhou 2002) will enable more flexible approach to supply chain management in mass customization.

3. MASS CUSTOMIZATION STRATEGIES

In order to fully understand the implications of a mass customization system, we must examine its characteristics and capabilities. Different companies adopt different methods or strategies for producing mass customized products. Mass customizers can be classified based on two main characteristics: the point of customer involvement in the design process and the type of modularity employed. Knowing the point of customer involvement and the modularity type are key elements in defining the configuration of processes and technologies that must be used to produce the mass customized product (Lampel and Mintzberg 1996).

The first identifier, point of customer involvement in the production cycle, is a key indicator of the degree of type of customization. Processes in which customer involvement is early in the cycle produce relatively more customized products than those with later customer involvement. Supporting this reasoning, we can view customization as taking one of three forms as described by Lampel and Mintzberg (1996).

1. A *pure* customized product is one that includes the customer throughout the entire production cycle. This strategy provides products that are completely unique to individual customer specifications.
2. A *tailored* customization strategy involves the customer at the point of fabrication. With this strategy, standard products are modified or altered to meet specific needs of a particular customer.
3. A *standardized* customization strategy involves the customer at the point of assembly and delivery. Here standard products are modified to customer specifications using a list of standardized options.

The second identifier, modularity type, is key to achieving cost effective customization by providing both economies of scale and scope. Modularity is the concept of decomposing a system into independent parts or modules that can be treated as logical units (Jiao and Tseng 2000). By using a modular approach, a product is designed such that part of the product is made in volume by standardized components. Customization is then achieved through a combination or modification of modules. A range of modularity types can be considered for different mass customizers. Ulrich and Tung (1991) have identified several types of modularity according to the way the product is assembled from modules:

1. Component sharing/swapping modularity.
2. Fabricate-to-fit modularity (dimensions of modules are changed).
3. Bus modularity (different modules can be attached to the common body).
4. Sectional modularity (different combinations of modules).

Salvador et al. (2002a) summarize a few other classifications of modularity according to the stability of the function allocated to the module criterion and the nature of their interface.

The concept of modularity distinguishes mass customization from pure customization by bounding the degree of customization that can be achieved, thus allowing repetitive manufacturing. When combining modularity along with customer involvement, customization can be fully realized in practice (Duray et al. 2000).

There are several other classifications of mass customization strategies, which have been summarized by Da Silveira et al. (2001) and MacCarthy et al. (2003). They mainly differ by the number of customization levels considered. MacCarthy et al. (2003) also criticize these classification approaches because these do not describe content of mass customization strategies. Authors propose a mode of operations concept for categorizing mass customization strategies. This concept differs from traditional classifications with its focus on operational characteristics of mass customization. Five modes proposed are catalogue, fixed resource design-per-order mass customization, flexible resource design-per-order mass customization, fixed resource call-off mass customization, and flexible resource call-off mass customization. Customization on a call-off basis (as opposed to once-only customization requests) implies that a manufacturer accepts a customization, if repeat orders are likely.

4. ISSUES IN SUPPLY CHAIN MANAGEMENT FOR MASS CUSTOMIZATION

As customer involvement is so relevant in adopting a mass customization strategy, a firm's communication with customers becomes essential. Additionally, suppliers also have an increasingly important role in the success of mass customization because of high requirements towards features of outsourced modules, which are used for customization. Unlike the traditional mass production systems, companies adopting this strategy must be able to efficiently produce, sort, ship and deliver small quantities of highly differentiated products (Berman 2002). Therefore, supply chain management is an essential piece of further progress of mass customization. It is argued that supply chain management encompasses other mass customization enablers, which also can exist at the single facility level (i.e., supply chain may involve multiple plants using advanced manufacturing technologies). Obviously, efficiency of supply chain management is highly dependent upon capabilities of communications infrastructure.

Differences between supply chain management in the mass customization framework and the mass production framework can be illustrated by flow of information and entities, respectively. Figures 1-1 and 1-2 illustrate the information flow and the roles that the customer and supplier play throughout the supply chain. As can be deduced from Figure 1-1, most functions in a mass production system are centralized. Inventory and production levels are internally determined based on sales forecasts. Customers are not directly involved in design or production planning and, therefore, do not have much contact with the firm. Inventory levels are replenished inside the firm's warehouse when the amount of material gets below a certain pre-determined safety stock. Since the firm can order replenishment before inventory fully depletes, quick and direct communication and distribution links between the firm and the supplier are not vital. Once production is complete, the finished product is stored in a finished goods warehouse, waiting for customers' orders. With the mass customization strategy (Figure 1-2), on the other hand, many of the important functions are decentralized. Production is directly based on customer specifications, making it essential for the firm to have an efficient means of dialogue with the customer. Because maintaining low inventory levels is one of the primary strategies with mass customizers, they must be able to procure materials from suppliers almost instantly. This need for high coordination makes it very important to have a quick and direct link with suppliers.

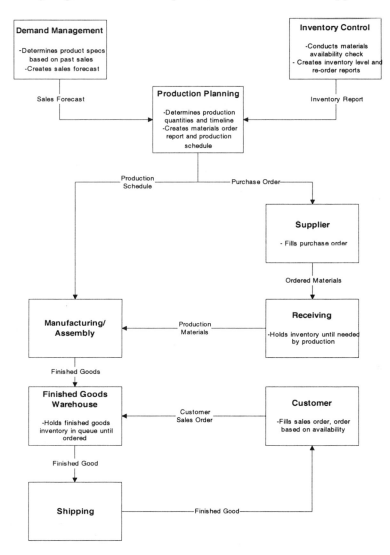

Figure 1-1. Information flow and entity relationships in the case of mass production

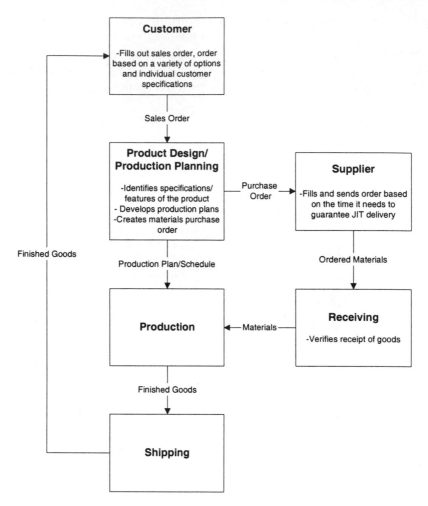

Figure 1-2. Information flow and entity relationships in the case of mass customization

Implementation of mass customization and supporting supply chain management functions cover a wide spectrum of problems and problem solving approaches, which are identified and described throughout the rest of the book. Some of the common systemic approaches to mass customization briefly introduced here are concurrent engineering, time-based manufacturing and postponement. Any of these approaches address specific stages of supply chain management for mass customization.

Concurrent engineering addresses problems of joint product and process design with participation of manufacturers, customers and suppliers. This synergy between design and manufacturing allows for easy manufacturing of products at a low product development cost (Tseng and Jiao 1998).

In order to support product customization, product family architecture fulfills customer needs by configuring and modifying well-established modules and components. Designing product families for mass customization involves identifying and capturing commonality among product designs and manufacturing processes and then using this commonality to develop versatile product platforms. Jiao and Tseng (2000) describe a product family architecture using three different perspectives to organize product varieties. These perspectives are the functional, the behavioral, and the structural. The functional perspective is seen from customer, sales and marketing viewpoint, the behavioral perspective from the product technology viewpoint, and the structural perspective from the manufacturing and logistics viewpoint.

Time-based manufacturing addresses the critical problem of proving lead times as short as possible. Time is treated as the strategic variable for establishing mass customization. This approach focuses on time compression techniques to enhance responsiveness to customer needs. Its main objective is to reduce end-to-end time to achieve quick responsiveness. Tu et al. (2001) identifies several sub-dimensions of time-based manufacturing processes that are important for the success of a mass customization system. These dimensions relate to shop floor employee involvement in problem solving, reengineering changeover processes, efficient preventive maintenance and reliability of suppliers.

Postponement mainly pertains to organization of distribution in the mass customization framework although it is also closely related to customer involvement and manufacturing issues. Postponement applied in the distribution channel allows for closer alignment with the customer and more rapid delivery. Through postponement, standard products can be made in the early stages of production and then later customized to customer order specifications at the point of distribution. This implies that materials management and some manufacturing activities must be done in the distribution channel. Mass customizers can outsource many of their materials management and manufacturing activities to third-party logistics services in order to realize efficient postponement (Van Hoek 2000).

5. CONCLUSIONS

As companies seek to improve their products and services to customers, mass customization has become a strategy of keen interest. Mass customization recognizes individual customer needs while achieving maximum reusability in order to maintain low production costs. This Chapter presents key attributes and factors necessary for the implementation of such system. Several strategies and techniques for managing a mass customization system are explored. Several industry applications are also presented in order to enhance understanding and illustrate the practical implications of mass customization.

REFERENCES

1. Aberdeen Group. Advanced Planning Engine Technologies: Can Capital Generating Technology Change the Face of Manufacturing? Aberdeen Group 1996; 2.
2. Aldanondo M., Hadj-Hamou K., Moynard G., Lamothe J. Mass customization and configuration: Requirement analysis and constraint based modeling propositions. Integrated Computer-Aided Engineering 2003; 10: 177.
3. Alford D., Sackett P., Nelder G. Mass customisation - an automotive perspective. International Journal of Production Economics 2000; 65: 99-110.
4. Berman B. Should your firm adopt a mass customization strategy? Business Horizons 2002; 45: 51-60.
5. Da Silveira G., Borenstein D., Fogliatto F. Mass customization: Literature review and research directions. International Journal of Production Economics 2001; 72: 1-13.
6. Duguay C. R., Landry S., Pasin F. From mass production to flexible/agile production. International Journal of Operations & Production Management 1997; 17: 1183-1195.
7. Duray R. M., Glen W. Improving customer satisfaction through mass customization. Quality Progress 1999; 32: 60-66.
8. Duray R., Ward P. T., Milligan G. W., Berry W. L. Approaches to mass customization: configurations and empirical validation. Journal of Operations Management 2000; 18: 605-625.
9. Duray R. Mass customization origins: mass or custom manufacturing? International Journal of Operations & Production Management 2002; 22: 314-328.
10. Eastwood M. A. Implementing mass customization. Computers in Industry 1996; 30: 171-174.
11. Feitzinger E., Lee H. L. Mass customization at Hewlett-Packard: The power of postponement. Harvard Business Review 1997; 75: 116-121.
12. Fisher M. L., Ittner C. D. The impact of product variety on automobile assembly operations: Empirical evidence and simulation analysis. Management Science 1999; 45: 771.
13. Fujita K. Product variety optimization under modular architecture. Computer-Aided Design 2002; 34: 953-965.
14. Gooley T. B. Mass customization: How Logistics makes it happen. Logistics 1998; 4: 49-53.
15. Griffiths J., Margetts D. Variation in production schedules - implications for both the company and its suppliers. Journal of Materials Processing Technology 2000; 103: 155-159.
16. Gunasekaran A., Goyal S. K., Martikainen T., Yli-Olli P. A conceptual framework for the implementation of zero inventory and just in time manufacturing concepts. Human Factors and Ergonomics in Manufacturing 1998; 8: 63-78.
17. Hart C. W. L. Mass customization: conceptual underpinnings, opportunities and limits. International Journal of Service Industry Management 1995; 6: 36-45.

18. Hayes R., Wheelwright S. C. Link manufacturing process and product life cycles. Harvard Business Review 1979; 57: 133-140.
19. Helander M. G., Jiao J. Research on E-product development (ePD) for mass customization. Technovation 2002; 22: 717-724.
20. Hirschhorn L., Noble P., Rankin T. Sociotechnical systems in an age of mass customization. Journal of Engineering and Technology Management 2001; 18: 241-252.
21. Huffman C., Kahn B. E. Variety for sale: Mass customization or mass confusion? Journal of Retailing 1998; 74: 491-513.
22. Jiao J., Tseng M. M. Fundamentals of product family architecture. Integrated Manufacturing Systems 2000; 11: 469-483.
23. Kotha S. From Mass Production to Mass Customization: The Case of the National Industrial Bicycle Company of Japan. European Management Journal 1996; 14: 442-450.
24. Lampel J., Mintzberg H. Customizing customization. Sloan Management Review 1996; 38: 21-30.
25. Lau R. S. M. Mass customization: the next industrial revolution. Industrial Management 1995; 37: 18-22.
26. MacCarthy B., Brabazon P. G., Bramham J. Fundamental modes of operation for mass customization. International Journal of Production Economics 2003; 85: 289-304.
27. Min H., Zhou G. Supply chain modeling: past, present and future. Computers & Industrial Engineering 2002; 43: 231-249.
28. Pine II B. J., Victor B., Boynton A. C. Making mass customization work. Harvard Business Review 1993; 108-119.
29. Poirier C. C., Reiter S. Supply Chain Optimization. San Francisco: Berrett-Koehler Publishers, 1996.
30. Salvador F., Forza C., Rungtusanatham M. Modularity, product variety, production volume, and component sourcing: theorizing beyond generic prescriptions. Journal of Operations Management 2002a; 20: 549-575.
31. Salvador F., Forza C., Rungtusanatham M. How to mass customize: Product architectures, sourcing configurations. Business Horizons 2002b; 45: 61-69.
32. Tseng M. M., Jiao J., Su C. J. Virtual prototyping for customized product development. Integrated Manufacturing Systems 1998; 9: 334-343.
33. Tseng M. M., Jiao J. Concurrent design for mass customization. Business Process Management Journal 1998; 4: 10-24.
34. Tu Q., Vonderembse M. A., Ragu-Nathan T. S. The impact of time-based manufacturing practices on mass customization and value to customer. Journal of Operations Management 2001; 19: 201-217.
35. Turbide D. The New World of Procurement. Midrange ERP. 1997; 12-16.
36. Turowski K. Agent-based e-commerce in case of mass customization. International Journal of Production Economics 2002; 75: 69-81.
37. Ulrich K., Tung K. Fundamentals of product modularity. Proceedings of the 1991 ASME Winter Annual Meeting Symposium on issues in Design/Manufacturing Integration 1991.
38. Van Hoek R. I. The role of third-party logistics providers in mass customization. The International Journal of Logistics Management 2000; 11: 37-46.

CHAPTER 2

CATEGORIZATION OF MASS CUSTOMZIATION RESEARCH

Charu Chandra[1], Jānis Grabis[2]

[1]University of Michigan-Dearborn
[2]Riga Technical University

Abstract: The categorization of mass customization research is compiled in order to provide an overall comprehension about research and application progress. The research articles are categorized using mass customization: strategy, enabler, analysis method and application area criteria. The overview results show that the standardized customization is dominant mass customization strategy, mass customization enablers are often investigated jointly and electronics is the most often-considered application area.

Keywords: Mass customization, research overview – mass customization.

1. INTRODUCTION

Implementation of mass customization involves solving many different problems, such as product design, supply chain management and production planning. This brings out the cross-sectional character of rapidly expanding research in the mass customization area. Additionally, mass customization appears to be an application driven subject making it appealing for applied studies.

There are many conceptual articles describing general principles of mass customization and related problems (see Pine et al. (1993) and Zipkin (2001) among others). Comprehensive overviews of the topic and analysis of literature are also compiled by Da Silveira et al. (2001) and Tseng and Jiao (2001). The primary objective of these articles, however, is not the analysis of individual problems. Additionally, each criterion is analyzed separately, making it difficult for identification of relationships among categorizations by different criteria.

In order to provide an overall comprehension about research and application progress in mass customization, a summary of relevant research articles is compiled in this chapter. This overview is compiled in the form of a table categorizing each article, considered according to several criteria along with short description of the article. This table can be used as a quick reference for finding articles dealing with some aspects of the mass customization problem area. The chapter also contains some summarized results of the complete overview.

The rest of the chapter is organized as follows. Section 2 describes the approach taken for research categorization. Section 3 gives overall results of the overview. Section 4 concludes. The completed overview table is placed in Appendix.

2. DESIGN

The survey of literature on mass customization is designed using often considered categorization criteria. The criteria used for categorization of articles are: mass customization strategy or level of mass customization considered, enablers used, analysis method used and application area. Additionally, keywords provided by authors (if these are not available, the main subject of a article is identified) are given and short description provided.

The level of mass customization criterion with some modifications is borrowed from Lampel and Mintzberg (1996) [(Da Silveira et al. (2001) provide a comprehensive discussion on different classifications for the level of mass customization]. Following levels of mass customization are considered: pure customization (P), tailored customization (T), customized standardization (S), additional custom work (A) and cosmetic (C). The term "cosmetic" introduced by Gilmore and Pine (1997) is used instead of Lampel and Mintzberg's segmented standardization because cosmetic is perceived as more general. Additional custom work level is included from Da Silveira et al. (2001), because it adds a transitional level between Lampel and Mintzberg's customized standardization and segmented standardization. This level includes both additional custom work and services as defined by Da Silveira et al. (2001). Enablers considered are agile manufacturing (A), supply chain management (S), customer driven design and manufacturing (C), lean manufacturing (L), advanced (manufacturing) technologies (T) and communications and networks (N) (from Da Silveira et al. 2001). The communication and networking enabler includes both hardware and software.

The analysis method criterion describes the investigation approach taken by respective authors. This property is important to provide an overview on methods and tools for dealing with mass customization problems. Methods are classified as mathematical modeling (MA), simulation (SI), survey (SU), conceptual (C), logical (L), modeling methodology (ME), application (A) and discussion (D). The modeling methodology method means that a particular problem solving methodology is developed. The conceptual method includes articles on identification of issues relevant to mass customization, classification of mass customization problems, review article, etc. Articles classified as discussion articles debate different mass customization problems. Logical methods include computerized models, and expert methods. The application area criterion is deemed important because it provides an idea about the level of penetration of mass customization in different industries.

In all 92 articles published in peer reviewed journals are surveyed. The list is aimed to cover the span of mass customization problem, although the objective is not to include every published article. The main emphasis is on articles dealing explicitly with mass customization. However, several articles from related areas such as, product design, postponement, make-to-order, just-in-time, flexibility, supporting information systems and third party logistics are also included.

3. RESULTS AND DISCUSSION

Complete results of the research categorization are given in Table 2-A1 in Appendix. This section gives the overall results.

Tables 2-1 through 2-3 categorize mass customization research according to mass customization - strategy, enabler and analysis method criteria, respectively. A majority of articles considering specific mass customization strategies investigate the standardized customization case. This suggests that modularization is probably well suited for achieving both high volume and customization at acceptable cost. Neither A nor C are frequently observed. Probably, analysis of these low customization levels already overlaps with mass production too tightly to make a distinction. Additionally, existence of different classifications for these levels of mass customization creates difficulties

All mass customization enablers considered are represented in the surveyed articles. Supply chain management and communications and networking are investigated more than others. Enablers, more frequently than other categorization criteria, are considered jointly. Communications and networking is frequently combined with supply chain management, agility and customer driven design. This is an obvious observation, since several factors contribute to the success of mass customization.

Given the relative novelty of the topic, there are many articles dealing with general and conceptual issues in mass customization. Application orientation is also a distinct characteristic pointing to unique characteristics of every mass customization adoption case. The presence of many logical models can be described by the importance of computerized support in adopting mass customization.

The application area by industries is the final categorization criteria. Adoption of mass customization is most often reported in the electronics sector. Automotive industry, bicycle production and industrial equipment are other areas mentioned more than twice. Some less traditional application areas such as education, healthcare and chemical industries are also reported.

Table 2-1. Categorization of articles according to mass customization strategy

Mass customization strategy	Articles
Pure customization (P)	Edwards 2002, Elliman and Orange 2003, Erens and Hegge, 1994, Istook 2002, Kotha 1996, Myung and Han 2001, Roy and Kodkani 1999, Tseng et al. 1998, Turowski 2002
Tailored customization (T)	Drejer and Gudmundsson 2002, Eastwood, 1996, Edwards 2002, Elliman and Orange 2003, Erens and Hegge, 1994, Hart 1996, Joneja and Lee 1998, Kotha 1996, Myung and Han 2001, Roy and Kodkani 1999, Storch and Sukapanpotharam 2003, Tseng and Jiao 1997, Tseng et al. 1998, Turowski 2002, Weng 1999, Yao and Carlson 2003
Customized standardization (S)	Aldanondo et al. 2003, Alfnes and Strandhagne 2000, Alford et al. 2000, Cheng et al. 2002, Fisher and Ittner 1999, Fujita 2002, Ghiassi and Spera 2003, Griffiths and Margetts 2000, Hart 1996, Karlsson 2002, Joneja and Lee 1998, Ma et al. 2002, Partanen and Haapasalo 2003, Randall and Ulrich 2001, Salvador et al. 2002a, Smirnov et al. 2003, Thompson and Nussbaum 2000, Tseng and Jiao 1997, Tseng and Jiao 1998, Yao and Carlson 2003, Wall 2000, Weng 1999
Additional custom work (A)	Rajagopal 2002, Salvador et al. 2002a, Twede et al. 2000, Van Hoek 2000,
Cosmetic (C)	Andel 2002, Daugherty et al. 1992, Feitzinger and Lee 1997, Fiang 2000, Rajagopal 2002, Salvador et al. 2002a, Twede et al. 2000, Van Hoek 2000

Table 2-2. Categorization of articles according to mass customization enablers

Enablers	Articles
Agile manufacturing (A)	Berman 2002, Duguay et al. 1997, Griffiths and Margetts 2000, Fulkerson 1997, Karsak and Kuzgunkaya 2002, Kim 1998, Penya et al. 2003, Worren et al. 2003, Yang and Li 2002, Yao and Carlson 2003
Supply chain management (S)	Akkermans, et al. 2003, Alford et al. 2000, Aviv and Federgruen 2001, Berman 2002, Cheng et al. 2002, Chiou et al. 2002, Contractor and Lorange 2002, Daugherty et al. 1992, Eastwood, 1996, Elliman and Orange 2003, Feitzinger and Lee 1997, Furst and Schmidt 2001, Ghiassi and Spera 2003, Gooley 1998, Griffiths and Margetts 2000, He and Jewkes 2000, Karlsson 2002, Ma et al. 2002, Partanen and Haapasalo 2003, Randall and Ulrich 2001, Saisse and Wilding 1997, Salvador et al. 2002a, Salvador et al. 2002b, Smirnov et al. 2003, Twede et al. 2000, Van Hoek 2000, Van Hoek 2001, Verwoerd 1999
Customer driven design and manufacturing (C)	Andel 2002, Bonney et al. 2003, Elliman and Orange 2003, Erens and Hegge, 1994, Helander and Jiao 2002, Istook 2002, Roy and Kodkani 1999, Tseng and Jiao 1997, Tseng et al. 1998
Lean manufacturing	Alfnes and Strandhagne 2000, Alford et al. 2000, Fisher and Ittner 1999, Griffiths and Margetts 2000, Hirschhorn et al. 2001, Partanen and Haapasalo 2003
Advanced technologies (T)	Alford et al. 2000, Bonney et al. 2003, Istook 2002, Eastwood, 1996, Edwards 2002, Elliman and Orange 2003, Fisher and Ittner 1999, Joneja and Lee 1998, Myung and Han 2001, Smirnov et al. 2003, Tseng and Jiao 1997, Tseng et al. 1998, Wall 2000
Communications and networks (N)	Akkermans, et al. 2003, Andel 2002, Contractor and Lorange 2002, Elliman and Orange 2003, Erens and Hegge, 1994, Fulkerson 1997, Furst and Schmidt 2001, Gardiner et al. 2002, Helander and Jiao 2002, Ghiassi and Spera 2003, Kohta 1996, Lee et al. 2000, Penya et al. 2003, Roy and Kodkani 1999, Sokolov 2001, Tseng et al. 1998, Turowski 2002, Walsh and Godfrey 2000, Weaver 2000, Yao and Carlson 2003

Table 2- 3. Categorization of articles according to analysis method

Analysis method	Articles
Discussion (D)	Andel 2002, Berman 2002, Contractor and Lorange 2002, Duguay et al. 1997, Eastwood, 1996, Elliman and Orange 2003, Gilmore and Pine 1997, Gooley 1998, Hart 1995, Hart 1996, Kakati 2002, Kotha 1996, Pine et al. 1993, Radder and Louw 1999, Sandelands 1994, Twede et al. 2000, Walsh and Godfrey 2000, Weaver 2000, Zipkin 2001
Conceptual (C)	Alfnes and Strandhagne 2000, Alford et al. 2000, Bonney et al. 2003, Da Silveira et al. 2001, Drejer and Gudmundsson 2002, Dubosson-Torbay et al. 2002, Duray et al. 2000, Edwards 2002, Fiang 2000, Fowler et al. 2000, Fujita 2002, Fulkerson 1997, Gilmore and Pine 1997, Kim 1998 Gunasekaran et al. 1998, Helander and Jiao 2002, Jiao et al., 2003, Karlsson 2002, Li and Li 2000, MacCarthy et al. 2003, Partanen and Haapasalo 2003, Saisse and Wilding 1997, Salvador et al. 2002b, Sokolov 2001, Svensson and Barfod 2002, Van Hoek 2001, Verwoerd 1999, Hirschhorn et al. 2001, Peters and Saidin 2000, Tseng and Jiao 1998, Wall 2000
Logical (L)	Akkermans, et al. 2003, Aldanondo et al. 2003, Erens and Hegge, 1994, Furst and Schmidt 2001, Gardiner et al. 2002, Ghiassi and Spera 2003, Myung and Han 2001, Penya et al. 2003, Roy and Kodkani 1999, Smirnov et al. 2003, Tseng et al. 1998, Turowski 2002, Yang and Li 2002
Survey (SU)	Duray et al. 2000, Kim 1998, Chiou et al. 2002, Daugherty et al. 1992, Duray 2002, Huffman and Kahn 1998, Rabinovich et al. 2003, Rajagopal 2002, Randall and Ulrich 2001, Thompson and Nussbaum 2000, Tu et al. 2001, Van Hoek 2000, Worren et al. 2003
Mathematical (MA)	Aviv and Federgruen 2001, Cheng et al. 2002, Fujita 2002, Gu et al. 2002, He and Jewkes 2000, Jiao and Tseng 2004, Karsak and Kuzgunkaya 2002, Lee et al. 2000, Weng 1999
Simulation (SI)	Fisher and Ittner 1999, Tseng et al. 1998
Methodology (ME)	Fulkerson 1997, Gu et al. 2002, Istook 2002, Joneja and Lee 1998, Storch and Sukapanpotharam 2003, Tseng and Jiao 1997, Tseng et al. 1998, Yao and Carlson 2003
Application (A)	Aviv and Federgruen 2001, Cheng et al. 2002, Drejer and Gudmundsson 2002, Eastwood 1996, Erens and Hegge 1994, Feitzinger and Lee 1997, Fisher and Ittner 1999, Fujita 2002, Gardiner et al. 2002, Griffiths and Margetts 2000, Gu et al. 2002, He and Jewkes 2000, Hirschhorn et al. 2001, Jiao and Tseng 2004, Karsak and Kuzgunkaya 2002, Kotha 1996, Lee et al. 2000, Ma et al. 2002, Peters and Saidin 2000, Salvador et al. 2002a, Tseng and Jiao 1998, Wall 2000, Weng 1999, Yao and Carlson 2003, Yang and Li 2002

APPENDIX

Table 2-A1 offers categorization of mass customization research. The Source column shows authors and year of the article. Articles are categorized according to mass customization: strategies, enablers, analysis method, and application area criteria (see Section 2 for description of criteria). Mass customization strategies used are pure customization (P), tailored customization (T), customized standardization (S), additional custom work (A) and cosmetic (C). Enablers used are agile manufacturing (A), supply chain management (S), customer driven design and manufacturing (C), lean manufacturing (L), advanced (manufacturing) technologies (T) and communications and networks (N). Analysis methods used are mathematical modeling (MA), simulation (SI), survey (SU), conceptual (C), logical (L), modeling methodology (ME), application (A) and discussion (D). If an article does not focus on any particular mass customization strategy, enabler or application area, or its focus is not clear, the corresponding table cells are marked with "-". Additionally, authors' keywords and a short description of the article are given.

Table 2-AI. Categorization of mass customization research

Source	MC strategy	Enablers	Keywords	Analysis method	Application area	Short description
Ahlstrom and Westbrook 1999	-	-	Flexibility, mass customization, operations management, surveys	SU	-	A survey of manufacturing companies that analyzes market changes driving customization, methods used to provide customized goods, positive and negative effects of customization, and difficulties of implementation.
Akkermans, et al. 2003	-	S N	Supply chain management, enterprise resource planning systems, Delphi study	L	-	Customization of products is forecasted to be one of the key issues in supply chain management. Enterprise Resource Planning systems are expected to substantially facilitate adoption of mass customization.
Aldanondo et al. 2003	S	-	Marketing, production management, configuration management, customization	L	-	An approach for identification of customization requirements and for evaluation of these requirements within the Constraint Satisfaction Problem is developed. The aim is to provide commercial configuration knowledge base designers with constraint-based generic modeling elements for customizable industrial product.
Alfnes and	S	L	Business	C	Furniture	A control model methodology is

Source	MC strategy	Enablers	Keywords	Analysis method	Application area	Short description
Strandhagne 2000			logistics, manufacturing processes			developed to design enterprises for mass customization. Principles of this methodology include differentiation of manufacturing, simplification of material flow, strategic positioning of stocks, decentralized decision-making in clearly defined control areas, and flow-oriented information.
Alford et al. 2000	S	S T L	Mass customization, automotive industry	C	Automotive	Types of customization in automotive industry are listed. Issues constraining customization and capabilities needed are discussed.
Andel 2002	C	C N	Packaging, Internet	D	Office products	General principles of mass customization are discussed.
Aviv and Federgruen 2001	-	S	Multi-item/ echelon/ stage inventory/ production, forecasting	MA	-	Postponement is analyzed under serially correlated demand.
Berman 2002	-	A S	Benefits of mass customization	D	-	Mass customization is compared with mass production. Potential benefits and components needed for successful implementation are discussed.
Bonney et al.	-	C T	Change,	C	-	Changes that occur in the

Source	MC strategy	Enablers	Keywords	Analysis method	Application area	Short description
2003			manufacturing companies, markets, relationship between production and inventory, concurrent engineering			manufacturing environment and their impact on production and inventory systems are identified. These include the product design process, reduction in product design time, new technology, increased competitiveness, need for shorter lead times, shorter product life cycles, the need for greater responsiveness. Response mechanisms to changes are discussed.
Cheng et al. 2002	S	S	Supply chain optimization, inventory/production, assembly systems, configure-to-order, postponement, Stochastic models	MA	Electronics	A product is configured from standardized components upon receiving a customer order. The configure-to-order system is argued to be an ideal approach to providing both mass customization and quick response. A base-stock policy for managing components inventory with service level requirements depending upon end-product configuration is developed.
Chiou et al. 2002	-	S	Postponement	SU	Electronics	Usage of postponement by Taiwan's information technology companies is analyzed. Postponement is shown to be

Source	MC strategy	Enablers	Keywords	Analysis method	Application area	Short description
						beneficial for products with high level of demanded customization and with modular design.
Contractor and Lorange 2002	-	S N	Alliances, knowledge	D	-	Creation of alliances and importance of knowledge are analyzed subject to different facilitating phenomena such as mass customization.
Da Silveira et al. 2001	-	-	Mass customization, customer-driven manufacturing, user involvement, agile manufacturing	C	-	Survey developments in mass customization, develop mass customization dimensions and indicate future research directions.
Daugherty et al. 1992	C	S	Responsiveness	SU	-	Ability of companies to accommodate custom distribution requests is analyzed.
Drejer and Gudmundsson 2002	T	-	Product development, new paradigm, case study	C A	Industrial Equipment	A multiple product development approach is proposed as a solution for building customized products at acceptable cost. It allows for streamlining internal production processes.
Dubosson-Torbay et al.	-	-	E-business models	C	-	Level of customization is identified as one of the factors for

Source	MC strategy	Enablers	Keywords	Analysis method	Application area	Short description
2002						classification of e-business models
Duguay et al. 1997	-	A	Overview, mass production, lean manufacturing, agile manufacturing	D	-	A historical account of the shift from mass production to flexible/agile production and a discussion of major aspects and differences between the two production systems are provided.
Duray 2002	-	-	Customization, Process planning, Manufacturing strategy	SU	-	Customizers are surveyed to follow evolution of mass producers and producers of crafted products to become mass customizers. The study shows plants that choose mass customization approaches matching the non-mass customized product line characteristics, have higher financial performance than those firms without a matched product line.
Duray et al. 2000	-	-	Mass customization, Process design, technology, Marketing/ operations interface	C SU	-	A conceptual typology of mass customization that provides an explicit means for identifying and categorizing mass customizers from the perspective of operations is developed. It suggests that two key variables in classifying mass customizers are the point in the production cycle, where the customer is involved in specifying

Source	MC strategy	Enablers	Keywords	Analysis method	Application area	Short description
						the product and the type of modularity used in the product.
Eastwood 1996	T	S T	Total customer satisfaction, mass customization	D A	Electronics	A comprehensive case study of the Motorola Company's implantation of mass customization for pagers, cell phones, and semiconductor.
Edwards 2002	P T	T	Design for manufacture, design for assembly, design for manufacture and assembly, new product development, concurrent engineering	C		Discussion of an important concurrent engineering imperative for cost effective product design, called Design for Manufacture and Assembly (DFMA). DFMA is a systemic approach for analyzing quantitative and qualitative product design information.
Elliman and Orange 2003	P T	S C T N	Distributed design, e-commerce, e-procurement, mass customization, process re-engineering, simulation modeling, supply chain	D	Construction	Growing demand for customization is analyzed. Electronic procurement is suggested as a response mechanism.

Source	MC strategy	Enablers	Keywords	Analysis method	Application area	Short description
Erens and Hegge 1994	P T	C N	Product design, information flow management	L A	Medical equipment	A product specification concept that allows a product specification both from customer/sales and manufacturing view.
Feitzinger and Lee 1997	C	S	Modularity, supply network	A	Electronics	Product differentiation for a specific customer is delayed until the latest possible point in the supply network. Product design and organizational capabilities are highlighted as main enablers of customization.
Fowler et al. 2002	-	-	Competencies, Knowledge, Dynamic Environment	C	-	A concept of dynamic competence to gain a competitive advantage, as compared to the product-centered strategy that many firms are adopting today. Dynamic competency strategy focuses on the consolidation of corporate wide technologies and production skills into competencies that allow businesses to quickly adapt to changing opportunities.
Fujita 2002	S	-	Product variety, design optimization, product family,	C MA	Aircraft and electronics	An optimization algorithm for selecting the level of product variety is developed. Modular architecture is used as a method for

Source	MC strategy	Enablers	Keywords	Analysis method	Application area	Short description
			modular architecture, mathematical modeling			achieving product variety.
Fulkerson 1997		A N	Mass customization, Assembly line sequencing	C ME	-	Importance of process flow management and enterprise resource planning systems for implementing mass customization is discussed. Application of genetic algorithms and agent systems is proposed.
Furst and Schmidt 2001	-	S N	Turbulent markets, extended enterprise, virtual enterprise, supply chain management, Extended Markup Language	L	Automotive	Mass customization is described as a driving force for organizational change and optimization, which leads to IT-support and virtual enterprises. Extended Markup Language based tools are developed to address these issues.
Gardiner et al. 2002	-	N	Enterprise Resource Planning systems, Data sharing, Cycle time reduction	L A	Electronics	An enterprise wide transaction structure such as Enterprise Resource Planning facilitates customization by improved cooperation between parties involved and reduced cycle times.

Source	MC strategy	Enablers	Keywords	Analysis method	Application area	Short description
Ghiassi and Spera 2003	S	S N	Synchronized supply chain, mass customization markets, e-commerce	L	Bicycle	A software system that can support business operations of a massively customized production system and supporting cooperation with supply chain partners is developed and its application is analyzed.
Gilmore and Pine 1997	-	-	General overview, types of customization	C D	-	General principles of mass customization, and types of customization are presented.
Gooley 1998	-	S	Types of customization, logistics	D	-	The concept of mass customization is discussed with emphasis on role of third party logistics providers.
Griffiths and Margetts 2000	S	A L S	Customer focused logistics, flexibility, production, mass customization	A	Automotive	A case study is used to analyze impact of adopting mass customization policies by a manufacturer on its suppliers.
Gu et al. 2002	-	-	Mass customization, optimization, systematic engineering	MA ME	Industrial equipment	Graphical models and mathematics models are used to describe various optimization methods for mass customization. The modeling scope includes implementation of mass customization, finding product families and determining

Source	MC strategy	Enablers	Keywords	Analysis method	Application area	Short description
						customization depth.
Gunasekaran et al. 1998	-	-	Inventory management, organizational readiness for MC	C	-	Critical issues related to implementation of concepts of zero inventory and just in time production are investigated. These concepts are often adopted to support mass customization. Behavioral and cultural aspects required by firms adapting zero inventory and just in time are also discussed.
Hart 1995	-	-	Mass customization, decision factors	D	-	General discussion on mass customization is provided. Key decision factors for adopting mass customization are analyzed.
Hart 1996	T S	-	Mass customization	D	Electronics, news, textile	Analyzing several manufacturing companies provides general discussion of mass customization.
He and Jewkes 2000	-	S	Make-to-order, queuing model	MA	-	Two algorithms for computing the average total cost per product and other performance measures for a make-to-order inventory-production system are developed. Algorithms are developed using matrix analytic methods.
Helander and Jiao 2002		C N	Global manufacturing;	C	-	Enablers and fundamental issues of applying the Internet to re-engineer

Source	MC strategy	Enablers	Keywords	Analysis method	Application area	Short description
			E-commerce, mass customization, product development, supply chain management			companies toward mass customization are identified and discussed.
Hirschhorn et al. 2001	-	L	STS, mass customization, learning organization autonomy	C A	Chemical industry	Linking mass customization to the task of building a learning organization, the case of redesign of a chemical pilot plant for producing new compounds is examined.
Huffman and Kahn 1998	-	-	Product variety, customer satisfaction	SU	Retail	Customer's response to variety offered is analyzed in order to determine their reaction on complexity inherent in wide variety of options. Customers are found to be more satisfied, if the choice is presented by attribute instead of alternative.
Istook 2002	P	C T	Manufacturing, customization, garments	ME	Textile	Activities involved in setting up the computer aided design system to automatically alter garments for individual fit are outlined.
Fiang 2000	C	-	Market segmentation,	C	-	Mass customization is discussed in relation to the market segmentation

Source	MC strategy	Enablers	Keywords	Analysis method	Application area	Short description
						theory. Ideas for developing a mass customization strategy are provided.
Jiao and Tseng 2004	-	-	Mass customization, product design, customizability, product and process platforms, performance measure, information content, process capability	MA	Power supply	Indices for determining product customizability, process customizability and perceived value of customization are developed.
Jiao et al., 2003	-	-	Mass customization product family, service delivery systems	C	-	Opportunities and challenges of mass customization for manufacturing industries and service providers are discussed. A technological road map for implementing mass customization based on building block identification, product platform development, and product life-cycle integration is proposed.
Joneja and Lee 1998	T S	T	Assembly, flexibility, modularity	ME	Industrial equipment	The modular, parametric assembly tool sets (MPATS) methodology developed is proposed to improve

Source	MC strategy	Enablers	Keywords	Analysis method	Application area	Short description
	-	-				setup times of assembly lines as well as lower tooling costs and speed up tooling lead times. Flexibility of assembly operations is effective in dealing with problems of mass customization.
Kakati 2002	-	-	Mass customization, modular design, customization strategy, flexibility, paint industry, and organization transformation	D	-	Innovation, quality improvement, time, flexibility, and cost management in every stage are essential pre-requisite of mass customization program. Measures for successful implementation of a mass customization strategy are suggested.
Karlsson 2002	S	S	Assembly, hybrid systems, customization	C	-	The concept of assembly-initiated production of customized products from standardized modules is discussed. Requirements for its implementation are outlined.
Karsak and Kuzgunkaya 2002	-	A	Multiple-objective decision making, fuzzy sets, investment decision analysis, flexible manufacturing	MA	-	Flexible manufacturing systems as an enabler of customization is analyzed. A model for selecting between multiple flexibility strategies is developed. It seeks for optimization of labor costs, work-in-process, setup costs, market response, quality, capital and

Source	MC strategy	Enablers	Keywords	Analysis method	Application area	Short description
			systems			maintenance cost, and floor space usage.
Kim 1998	-	A	Intranet, virtual corporation, analytic hierarchy process	C SU	-	An analytical structure of Intranet functions in relation to building virtual organizations is constructed. The relative importance of these functions is partially attributed to the impact of mass customization.
Kotha 1996	P T	N	Implementation of MC, success factors	D A	Bicycle	A detailed discussion on implementation of mass customization for bicycle production is provided. Interactions between mass customization and mass production are analyzed.
Lee et al. 2000	-	N	Complementarity, electronic commerce, mass customization	MA	-	Two business strategies, electronic commerce and mass customization, in a profit maximization model are investigated. These strategies are shown as complementary under certain assumptions.
Li and Li 2000	-	-	Systems science, systems approach, manufacturing,	C	-	Mass customization is identified as one of the trends calling for a more integrated approach to manufacturing. An integrated

Source	MC strategy	Enablers	Keywords	Analysis method	Application area	Short description
			manufacturing information systems, production and operations management			information systems approach based on the general systems concept has been suggested to meet the challenge derived from new trends.
Ma et al. 2002	S	S	Component commonality, manufacturing postponement, fill rates, multistage assembly systems	A	-	A model for location of the decoupling point in multi-stage systems is developed. Interactions between processing and procurement lead times are a major factor affecting the decision.
MacCarthy et al. 2003	-	-	Mass customization, taxonomy	C	-	Mass customization classification schemas are reviewed. The taxonomy of operational modes for mass customization is developed on the basis of case studies.
Fisher and Ittner 1999	S	T L	Product variety, labor productivity, product complexity, lean production, automotive assembly, mass production,	SI A	Automotive	The automotive assembly process including installation of custom options is analyzed. Improved manufacturing technology allowing for shorter setup times and manufacturing flexibility are mentioned as main operational enablers of customization. Just-in-time deliveries are argued not only

Source	MC strategy	Enablers	Keywords	Analysis method	Application area	Short description
			flexible production			to reduce inventory, but also to speedup the assembly process by delivering components in the right assembly order.
Myung and Han 2001	P T	T	Functional feature, design unit, configuration design, parametric modeling, design expert system	L	Machine tools	Importance of parametric modeling and configuration design methods for mass customization is emphasized. A framework for a system, composed of commercial computer aided design system and an expert system, which can parametrically model both parts and assemblies of a product, is developed.
Partanen and Haapasalo 2003	S	S L	Fast production, mass customization, modularity, customer focus	C	Electronics	The concept of fast production is discussed.
Penya et al. 2003	-	A N	Business planning, enterprise resource planning, marketing, manufacturing processes, customization	L	-	The PABADIS project in progress is aimed to achieve an intelligent manufacturing system with a product-oriented approach, suitable primarily for single-piece production systems, but also for the challenges of mass-customized production.

Source	MC strategy	Enablers	Keywords	Analysis method	Application area	Short description
Peters and Saidin 2000	-	-	Information technology, information marketing, mass customization, services marketing	C A	Computers	Challenge of implementing mass customization in a services context is explored. Drivers and problem issues are identified with emphasis on prioritizing of problems.
Pine et al. 1993	-	-	Mass customization, success factors	D	-	Importance of keeping drive for customization focused on value achieved is highlighted. Success factors for implementation of mass customization are discussed.
Rabinovich et al. 2003	-	-	Logistics/distribution, MIS/operations interface, empirical research methods, structural equation modeling	SU	-	Mass customization is shown to have positive impact on end-product inventory.
Radder and Louw 1999	-	-	Customer satisfaction, Mass customization, Mass production	D	-	Differences and touch points between mass production and mass customization are discussed.

Source	MC strategy	Enablers	Keywords	Analysis method	Application area	Short description
Rajagopal 2002	A C	-	Enterprise resource planning systems, process model, causal model, contextual factors, triangulation, integration and performance	SU	-	Implementation of Enterprise Resource Planning systems is studied. Customization of these systems is described as one of implementation steps.
Randall and Ulrich 2001	S	S	Product variety, supply chain structure, firm performance	SU	Bicycle	Relationships between product variety, supply chain structure and performance of companies are analyzed. No evidence has been found that offering more variety through strategies of mass customization or variety postponement results in higher firm performance. However, authors indicate that this finding does not call for general dismissal of such strategies.
Roy and Kodkani 1999	P T	C N	Product design, virtual enterprise	L	Industrial equipment	A prototype system is developed to aid product development teams to perform their computer aided design activities within the collaborative framework of

Source	MC strategy	Enablers	Keywords	Analysis method	Application area	Short description
						Internet technologies.
Saisse and Wilding 1997	-	S	Short-term decisions	C	-	The concept of short-term strategic management is discussed. Its utilization for development of manufacturing management tools for mass customization industries is analyzed.
Salvador et al. 2002a	S A C	S	Case study research, product variety, sourcing, modularity, supply chain management, operational performance	A	Automotive, communications, equipment, food processing equipment	Multiple case studies are analyzed to identify how manufacturing characteristics affect the appropriate type of modularity to be embedded into the product family architecture, and how types of modularity relate to component sourcing.
Salvador et al. 2002b	-	S	Product and process design, supply chain	C	-	Factors facilitating growing importance of mass customization are analyzed. Requirements to product, process and supply chain design are discussed.
Sandelands 1994	-	-	Mass customization	D	-	Mass customization is advocated as a new business model atop of previous ones.
Smirnov et al. 2003	S	T S	Agents, data processing, constraints,	L	Knowledge management	A knowledge source network approach is proposed for customization of knowledge

Source	MC strategy	Enablers	Keywords	Analysis method	Application area	Short description
			configuration management			management.
Sokolov 2001	-	N	Customized education, individual learning, standard study interface, two tier classroom, student focused organization	C	Education	The idea of mass-customization in the educational system is introduced. Customized education is an education system in which technologies and organizational skills are combined to provide for the individual's educational needs, when and where they are required.
Storch and Sukapanpothara m 2003	T	-	Ship design, design for production, mass customization	ME	Ship building	Theory and methodology for developing type blocks for ship design are developed. These blocks allow for efficient design of customized products.
Svensson and Barfod 2002	-	-	Mass customization, industrialization, Small and Medium Enterprises	C	-	Industrial level implementation of mass customization in Small and Medium Enterprises is investigated. A procedure for implementation of mass customization is provided.
Thompson and Nussbaum 2000	S	-	Mass customization, healthcare	SU	Healthcare	A survey is conducted (and its results analyzed and reported) in order to identify customer preferences and subsequently to develop customized services.

Source	MC strategy	Enablers	Keywords	Analysis method	Application area	Short description
Tseng and Jiao 1997	T S	C T	Case-based design, mass customization, product design, object-oriented, design automation	ME	Power supply	Reuse of design elements implemented using case based reasoning is advocated as an approach to designing customized products. The implementation of case based reasoning in the mass customization context is discussed
Tseng and Jiao 1998	S	-	Concurrent engineering	C A	Electronics	The role of concurrent product design in mass customization is discussed. An approach for systematic development of product families, optimizing match between customer requirements and manufacturing capabilities is proposed.
Tseng et al. 1998	P T	C T N	Design environment, prototyping, simulation	ME L SI	Consumer products	Virtual prototyping is combined with design using simulation so that customization requirements and process capabilities of a firm can be balanced early in the design phase.
Tu et al. 2001	-	-	Mass customization, time-based manufacturing, value to customers, structural	SU	-	A framework for understanding relationships among time-based manufacturing practices, mass customization, and value to the customer is developed. The study indicates that firms with high levels of time-based manufacturing

Source	MC strategy	Enablers	Keywords	Analysis method	Application area	Short description
			equation modeling			practices have high levels of mass customization and value to the customer.
Turowski 2002	P T	N	E-commerce, mass customization, software agent, Extended Markup Language, Electronic Data Interchange	L	.	The role of the Internet and Extended Markup Language based Electronic Data Interchange in conveying custom requirements from customer to producer is analyzed.
Twede et al. 2000	A C	S	Postponement, mass customization, international distribution, packaging	D	-	Postponement applications have been growing in recent years as international firms increasingly develop global products that are customized for local markets. Logistical principles of postponement and speculation are outlined. Factors favouring postponement are explored.
Van Hoek 2000	A C	S	Third party logistics	SU	-	A survey about expansion of third-party logistics services into areas of postponement and customization is conducted.
Van Hoek 2001	-	S	Logistics/ distribution,	C	-	Postponement is discussed as a method for delaying activities in

Source	MC strategy	Enablers	Keywords	Analysis method	Application area	Short description
	-		flexible manufacturing systems, operations strategy			the supply chain until customer orders are received with the intention of customizing products. Relationships between postponement and the level of mass customization are analyzed.
Vewoerd 1999		S	Value added logistics	C	Electronics	Customization possibilities at various stages of the supply chain, inventory considerations, speed of production processes, speed of distribution processes and speed of information processing are mentioned among factors influencing location of the decoupling point for postponement.
Wall 2000	S	T	Evaluation of customization opportunities	C A	Telecommunications	Describes an approach taken by the British Telecom to explore opportunities for adopting mass customization.
Walsh and Godfrey 2000	-	N	E-commerce	D	Retail	E-commerce is described as a way to provide customized retail services to customers. It also facilitates distribution of customized products.
Weaver 2000	-	N	Composites, e-commerce	D	Chemical	Demand for customized products boosts performance of

Source	MC strategy	Enablers	Keywords	Analysis method	Application area	Short description
Weaver 2000	-	N	Composites, e-commerce	D	Chemical	Demand for customized products boosts performance of manufacturers of composite materials.
Weng 1999	T S	-	Strategies for MC, application, make-to-order manufacturing	MA	Lighting fixture	A real world make-to-order system is described. A model for studying the conflicting objectives between delivery time and profitability is developed. The model analysis highlights significant impact of the order acceptance rate on profitability.
Worren et al. 2003	-	A	Strategic flexibility, modularity, global strategies	SU	-	Assuming that modular product architectures are key enablers to strategy flexibility, an integrative conceptual model encompassing the contributing factors and outcomes of modularity is developed. The model indicates that variety is positively related to firm performance and product modularity is positively related to model variety.
Yang and Li 2002	-	A	Mass customization, agile manufacturing, agility	L A	Casting	A systematic method for evaluating the agility of a mass customization system is established. An application example is discussed.

Source	MC strategy	Enablers	Keywords	Analysis method	Application area	Short description
Yang and Li 2002	-	A	Mass customization, agile manufacturing, agility evaluation	L A	Casting	A systematic method for evaluating the agility of a mass customization system is established. An application example is discussed.
Yao and Carlson 2003	T S	A N	Agile manufacturing, lean manufacturing, mass customization, automatic data collection	ME A	Furniture industry	Describes a decision support system for managing an agile manufacturing facility for production of customized low cost products.
Zipkin 2001	-	-	Customization, marketing	D	-	Mass customization is presented as one of the possible competitive strategy. A need for matching demand for customization and the level of customization offered is emphasized. Factors limiting implementation of mass customization are outlined.

LEGEND.

MC strategy column: *P- pure customization, T - tailored customization, S - customized standardization, A - additional custom work and C - cosmetic.*

Enablers column: *A - agile manufacturing, S - supply chain management, C - customer driven design and manufacturing, L – lean manufacturing, T - advanced (manufacturing) technologies and N - communications and networks.*

Analysis method column: *MA - mathematical modeling, SI - simulation, SU - survey, C - conceptual, L - logical, ME – modeling methodology, A -application and D – discussion.*

REFERENCES

1. Ahlstrom P., Westbrook R. Implications of mass customization for operations management - An exploratory survey. International Journal of Operations & Production Management 1999; 19: 262-274.
2. Akkermans H. A., Bogerd P., Yucesan E., van Wassenhove L. N. The impact of ERP on supply chain management: Exploratory findings from a European Delphi study. European Journal of Operational Research 2003; 146: 284-301.
3. Aldanondo M., Hadj-Hamou K., Moynard G., Lamothe J. Mass customization and configuration: Requirement analysis and constraint based modeling propositions. Integrated Computer-Aided Engineering 2003; 10: 177.
4. Alfnes E., Strandhagen J. O. Enterprise Design for Mass Customisation: The Control Model Methodology. International Journal of Logistics: Research and Applications 2000; 3: 111-125.
5. Alford D., Sackett P., Nelder G. Mass customisation - an automotive perspective. International Journal of Production Economics 2000; 65: 99-110.
6. Andel T. From common to custom: The case for make-to-order. Material Handling Management 2002; 24-31.
7. Aviv Y., Federgruen A. Design for postponement: A comprehensive characterization of its benefits under unknown demand distribution. Operations Research 2001; 49: 578-598.
8. Berman B. Should your firm adopt a mass customization strategy? Business Horizons 2002; 45: 51-60.
9. Bonney M., Ratchev S., Moualek I. The changing relationship between production and inventory examined in a concurrent engineering context. International Journal of Production Economics 2003; 81-82: 243-254.
10. Cheng F., Ettl M., Lin G., Yao D. D. Inventory-service optimization in configure-to-order systems. Manufacturing & Service Operations Management 2002; 4: 114-132.
11. Chiou J.-S., Wu L.-Y., Hsy J. C. The adoption of form postponement strategy in a global logistics system: The case of Taiwanese information technology industry. Journal of Business Logistics 2002; 23: 107-124.
12. Contractor F. J., Lorange P. The growth of alliances in knowledge-based economy. International Business Review 2002; 11: 485-502.
13. Da Silveira G., Borenstein D., Fogliatto F. Mass customization: Literature review and research directions. International Journal of Production Economics 2001; 72: 1-13.
14. Daugherty P. J., Sabath R. E., Rogers D. S. Competitive advantage through customer responsiveness. Logistics and Transportation Review 1992; 28: 257-271.
15. Drejer A., Gudmundsson A. Towards multiple product development. Technovation 2002; 22: 733-745.
16. Dubosson-Torbay M., Osterwalder A., Pigneur Y. E-business model design, classification, and measurements. Thunderbird International Business Review 2002; 44: 5-23.
17. Duguay C. R., Landry S., Pasin F. From mass production to flexible/agile production. International Journal of Operations & Production Management 1997; 17: 1183-1195.
18. Duray R., Ward P. T., Milligan G. W., Berry W. L. Approaches to mass customization: configurations and empirical validation. Journal of Operations Management 2000; 18: 605-625.
19. Duray R. Mass customization origins: mass or custom manufacturing? International Journal of Operations & Production Management 2002; 22: 314-328.

20. Eastwood M. A. Implementing mass customization. Computers in Industry 1996; 30: 171-174.
21. Edwards K. L. Towards more strategic product design for manufacture and assembly: priorities for concurrent engineering. Materials and Design 2002; 23: 651-656.
22. Elliman T., Orange G. Developing distributed design capabilities in the construction supply chain. Construction Innovation 2003; 3: 15.
23. Erens F. J., Hegge H. M. H. Manufacturing and sales co-ordination for product variety. International Journal of Production Economics 1994; 37: 83-99.
24. Feitzinger E., Lee H. L. Mass customization at Hewlett-Packard: The power of postponement. Harvard Business Review 1997; 75: 116-121.
25. Fiang P. Segment based mass customization: an exploration of a new conceptual marketing framework. Internet Research: Electronic Networking Applications and Policy 2000; 10: 215-226.
26. Fisher M. L., Ittner C. D. The impact of product variety on automobile assembly operations: Empirical evidence and simulation analysis. Management Science 1999; 45: 771.
27. Fowler S. W., King A. W., Marsh S. J., Victor B. Beyond products: new strategic imperatives for developing competencies in dynamic environments. Journal of Engineering and Technology Management 2000; 17: 357-377.
28. Fujita K. Product variety optimization under modular architecture. Computer-Aided Design 2002; 34: 953-965.
29. Fulkerson F. A response to dynamic change in the market place. Decision Support Systems 1997; 21: 199-214.
30. Furst K., Schmidt T. Turbulent markets need flexible supply chain communication. Production Planning & Control 2001; 12: 525-533.
31. Gardiner S. C., Hanna J. B., LaTour M. S. ERP and the reengineering of industrial marketing processes: A prescriptive overview for the new-age marketing manager. Industrial Marketing Management 2002; 31: 357-365.
32. Ghiassi M., Spera C. Defining the Internet-based supply chain system for mass customized markets. Computers & Industrial Engineering 2003; 45: 17-41.
33. Gilmore J. H., Pine B. J. The four faces of mass customization. Harvard Business Review 1997; 75: 91-102.
34. Gooley T. B. Mass customization: How Logistics makes it happen. Logistics 1998; 4: 49-53.
35. Griffiths J., Margetts D. Variation in production schedules -- implications for both the company and its suppliers. Journal of Materials Processing Technology 2000; 103: 155-159.
36. Gu X. J., Qi G. N., Yang Z. X., Zheng G. J. Research of the optimization methods for mass customization (MC). Journal of Materials Processing Technology 2002; 129: 507-512.
37. Gunasekaran A., Goyal S. K., Martikainen T., Yli-Olli P. A conceptual framework for the implementation of zero inventory and just in time manufacturing concepts. Human Factors and Ergonomics in Manufacturing 1998; 8: 63-78.
38. Hart C. W. L. Mass customization: conceptual underpinnings, opportunities and limits. International Journal of Service Industry Management 1995; 6: 36-45.
39. Hart C. W. Made to order. Marketing Management 1996; 5: 12-22.
40. He Q.-M., Jewkes E. M. Performance measures of a make-to-order inventory-production system. IIE Transactions 2000; 32: 409-419.
41. Helander M. G., Jiao J. Research on E-product development (ePD) for mass customization. Technovation 2002; 22: 717-724.

42. Hirschhorn L., Noble P., Rankin T. Sociotechnical systems in an age of mass customization. Journal of Engineering and Technology Management 2001; 18: 241-252.
43. Huffman C., Kahn B. E. Variety for sale: Mass customization or mass confusion? Journal of Retailing 1998; 74: 491-513.
44. Istook C. L. Enabling mass customization: computer-driven alteration methods. International Journal of Clothing 2002; 14: 61-76.
45. Jiao J., Tseng M. M. Customizability analysis in design for mass customization. Computer-Aided Design, In press.
46. Jiao J., Ma Q., Tseng M. M. Towards high value-added products and services: mass customization and beyond. Technovation 2003; 23: 809-821.
47. Joneja A., Lee N. A modular, parametric virbratory feeder: a case study for flexible assembly tools for mass customization. IIE Transactions 1998; 30: 923-931.
48. Kakati M. Mass customization – needs to go beyond technology. Human Systems Management 2002; 21: 85–93.
49. Karlsson A. Assembly-initiated production - a strategy for mass-customization utilizating modular, hybrid automatic production systems. Assembly Automation 2002; 22: 239-247.
50. Karsak E. E., Kuzgunkaya O. A fuzzy mutliple objective programming approach for the selection of flexible manufacturing system. International Journal of Production Economics 2002; 79: 101-111.
51. Kim J. Hierarchical structure of intranet functions and their relative importance: using the analytic hierarchy process for virtual organization. Decision Support Systems 1998; 23: 59-74.
52. Kotha S. From Mass Production to Mass Customization: The Case of the National Industrial Bicycle Company of Japan. European Management Journal 1996; 14: 442-450.
53. Lampel J., Mintzberg H. Customizing customization. Sloan Management Review 1996; 38: 21-30.
54. Lee C.-H., Barua S., Whinston A., B. A. The complementarity of mass customization and electronic commerce. Economics of Innovation & New Technology 2000; 9: 81.
55. Li H., Li L. X. Integrating systems concept into manufacturing information systems. Systems Research and Behavioral Science 2000; 17: 135-147.
56. Ma S., Wang W., Liu L. Commonality and postponement in multistage assembly systems. European Journal of Operational Research 2002; 142: 523-538.
57. MacCarthy B., Brabazon P. G., Bramham J. Fundamental modes of operation for mass customization. International Journal of Production Economics 2003; 85: 289-304.
58. Myung S., Han S. Knowledge-based parametric design of mechanical products based on configuration design methods. Expert Systems with Applications 2001; 21: 99-107.
59. Partanen J., Haapasalo H. Fast production for order fulfillment: Implementing mass customization in electronics industry. International Journal of Production Economics, In press.
60. Penya Y. K., Bratoukhine A., Sauter T. Agent-driven distributed-manufacturing model for mass customisation. Integrated Computer-Aided Engineering 2003; 10: 139.
61. Peters L., Saidin H. IT and the mass customization of services: the challenge of implementation. International Journal of Information Management 2000; 20: 103-119.
62. Pine II B. J., Victor B., Boynton A. C. Making mass customization work. Harvard Business Review 1993; 108-119.
63. Rabinovich E., Dresner M. E., Evers P. T. Assessing the effects of operational processes and information systems on inventory performance. Journal of Operations Management 2003; 21: 63-80.
64. Radder L., Louw L. Mass customization and mass production. The TQM Magazine 1999; 11: 35-40.

SECTION 2:

PROBLEM SOLVING FRAMEWORKS, MODELS, AND METHODOLOGIES

CHAPTER 3

MASS CUSTOMIZATION: FRAMEWORK AND METHODOLOGIES

Charu Chandra[1], Jānis Grabis[2]

[1]*University of Michigan-Dearborn*
[2]*Riga Technical University*

Abstract: This Chapter places mass customization in the wider perspective of visionary evolution of manufacturing systems. Reconfigurable products and processes that enable adapting to changing market conditions are central to advanced manufacturing systems, including implementation of mass customization. Trans-organizational character of modern manufacturing systems calls for analyzing reconfiguration issues from the supply chain perspective. Problem-solving approaches for managing reconfigurable supply chains are discussed. A decision support system integrating the information support system and the decision modeling system is proposed.

Keywords: Reconfigurable system, supply chain reconfiguration, decision support system.

1. INTRODUCTION

Mass customization strategies have the potential of being one of the major catalysts to achieve *Manufacturing Visions* identified by the manufacturing community at large (National Research Council 1998). The key technologies to nurture in this endeavor are:

- Adaptable, and integrated equipment, processes, and systems that can be readily reconfigured,
- Manufacturing processes that minimize waste,
- System synthesis, modeling, and simulation for all manufacturing operations,
- Technologies to convert information into knowledge for effective decision making,
- Software for intelligent collaboration systems, and
- New educational and training methods that enable the rapid assimilation of knowledge.

The common thread in the deployment of these technologies is achieving (a) reconfigurability, (b) efficiency, and (c) complex modeling and analysis in decision-making related to managing advanced manufacturing systems.

This emphasis on developing enhanced manufacturing capabilities and technologies to support its infrastructure, mandates research in following crosscutting areas:

- Adaptable and reconfigurable manufacturing systems.
- Information and communication technologies.
  - Processes for capturing and using knowledge for manufacturing.
  - IT adopted and incorporated into collaboration systems and models focused on improving methods for people to make decisions, individually and as a group.
- Enterprise modeling and simulation.
  - Analytical tools for modeling and assessment.
    - Managing and using information to make intelligent decisions among a vast array of alternatives.
    - Adapting and reconfiguring manufacturing enterprises to enable formation of complex alliances with other organizations.

The objective of this Chapter is to exploit the rich diversity of knowledge acquired from the study of these crosscutting research areas in, (a) laying the foundation of a mass customization framework in relation to reconfigurable manufacturing systems, (b) to survey existing methodological approaches to mass customization, and (c) to describe integration of qualitative and quantitative modeling approaches enabling efficient problem solving. Laying the foundations of a reconfigurable system with dimensions of manufacturing, levels of mass customization, and supporting logistics can best achieve this objective.

The rest of the Chapter is organized as follows. The mass customization framework in relation to reconfigurable systems is analyzed in Section 2. In Section 3, the nature of reconfiguration for Manufacturing Systems is described. Triggers and impacts that a reconfigurable supply chain (SC) required supporting these manufacturing systems are identified. Section 4 provides an overview of above-mentioned inter-disciplinary research to problem solving for these complex systems. An integrated modeling approach to problem solving is presented in Section 5. Section 6 concludes.

2. RECONFIGURATION AND MASS CUSTOMIZATION

Reconfigurable manufacturing systems can be designed, modeled, and configured for specific applications, and upgraded and reconfigured rather than replaced. With a reconfigurable system, new products and processes can be introduced with considerably less expense and ramp-up time.

The reconfigurability property of an advanced manufacturing system, described at the beginning of this Chapter is one of the principal properties, because other properties either support reconfigurability or must be present in the efficient reconfigurable system. Minimization of waste, although more difficult to achieve, is as important in the reconfigurable system, as in a more rigid system. Reconfiguration efficiency can be achieved only by means of intelligent decision making (i.e., use of system synthesis, analysis and simulation). Knowledge is essential for predicting and assessing impact of changes due to reconfiguration on the manufacturing system. Changes in the system must be supported by information systems that enable all parties involved to learn about these changes and adjust their processes. Success of reconfiguration also depends upon decision-maker's ability to adapt to the modified system based on knowledge acquired through education and training.

Implementation of mass customization heavily draws upon reconfigurability and the associated properties of the advanced manufacturing system. In other words, if an enterprise has been successful in implementing advanced manufacturing systems, it is also capable of producing mass customized products. Figure 3-1 categorizes selected mass customization research articles according to problem areas in advanced manufacturing system design and implementation. It demonstrates that many issues in implementing mass customization are directly related to the drive for the advanced manufacturing system. The reconfigurable system, waste minimization and software for collaboration properties can be related to mass customization enablers (Da Silveira et al. 2001) such as agility, supply chain management, advanced technologies, lean manufacturing and networking. Other properties, including synthesis and modeling, knowledge utilization and training provide more general framework and pertain to all enablers.

One issue of major importance for mass customization, not explicitly addressed in the advanced manufacturing system context is customer involvement. This issue interacts with the reconfigurable system property through customers' ability to reconfigure products often achieved by means of modular product design (e.g., Salvador et al. 2002a). Customer requirements influence the way manufacturers reconfigure manufacturing processes (Tseng et al. 1998), and minimization of waste is one of their primary objectives. Synthesis and modeling of systems providing mass customized products depends upon the level of customization required by customers and manufacturer's ability to meet these requirements (Gu et al. 2002). Information support systems are essential for effective cooperation between customers and manufacturers, and further channeling customer requirements up along the SC (Furst and Schimdt 2001).

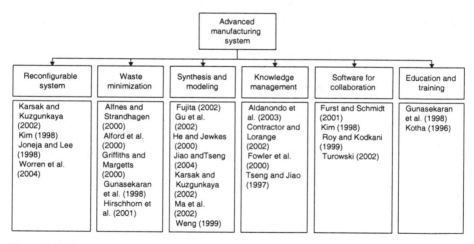

Figure 3-1. Categorization of selected mass customization research in advanced manufacturing system

3. RECONFIGURABLE SUPPLY CHAINS

A reconfigurable SC is a system that efficiently adapts to its environment, offered in the form of supply and demand issues for the product(s) to be manufactured. A reconfigurable SC is needed to manage logistics in a reconfigurable manufacturing system. This is because the adopted policies for product, process, and resource components of a reconfigurable manufacturing system have to be integrated with both inbound and outbound logistics decision, in order to realize benefits of mass customization strategies. Some of the key triggers for designing and implementing reconfigurable SC are as follows:

- Introduction of new product(s), or upgrade for existing product(s),
- Introduction of new, or improvement in existing process(es),
- Allocation of new, or re-allocation of existing resource(s),
- Selection of new supplier(s), or de-selection of existing ones,
- Changes in demand patterns for product(s) manufactured,
- Changes in lead times for product and / or process life cycles, and
- Changes in commitments within or between SC members.

A reconfigurable SC can help in assessing impacts of one or more of the following factors / activities in a reconfigurable manufacturing system:

- Flows due to materials, inventory, information, and cash.
- Throughput due to movement of product.
- Capacity utilization.
- Costs at various stages of product development life cycle.
- Lead time in product development.
- Batch and lot sizing.
- Process redesign.
- Product development strategies.
- Procurement and / or allocation of resources.
- Strategic, tactical, and operational policies for the SC.

Analysis of these factors/activities involves dealing with wide range of managerial problems and spans across all tiers of the SC. Problem-solving approaches need to consider both interactions among factors and activities, and SC members.

4. PROBLEM-SOLVING FOR RECONFIGURABLE SYSTEMS

4.1 State-of the-research

In order to provide an integrated overview of interconnectedness of cross cutting research areas for reconfigurable systems, three problem-solving approaches are proposed: systemic, reductionist, and analytic. These are defined as follows:

1. *Systemic Approach* incorporates the abstract level. This "level of inquiry" deals with issues of scalability of system, meta-modeling of systems, and defining the dynamic knowledge problem domain model.
2. *Reductionist Approach* incorporates the activity level. This "level of inquiry" consists of dynamic knowledge problem domain model, internal state, and goals and objectives of the enterprise *units*, e.g., producer, plant, department, supplier, vendor, etc., and strategic management models.
3. *Analytic Approach* incorporates the implementation level. This "level of inquiry" consists of internal state, goals and objectives of the enterprise units, strategic management models, and shared goals and objectives of the enterprise.

Related research in direct comparison to the three problem-solving approaches described above is discussed next.

At the *systemic* level, a SC is a general class of System that exhibits a cooperative behavior within its business and market environment (Klir 1991). The foundation of this system is built on a network architecture that has various demand and supply nodes as it provides, as well receives goods and services to and from its customers and suppliers, respectively (Chandra 1997, Lee and Billington 1993, Swaminathan et al. 1998). SC systems framework describes general foundational elements of integration between its marketing and production functions. These are in the form of general theories, hypotheses, standards, procedures, and models that are based on well-founded principles in these disciplines (Cohen and Lee 1988, 1989, Deleersnyder et al. 1992, Diks and Kok 1998, Drew 1975, Graves 1982, Hackman and Leachman 1989, Lee 1996, NIST 1999, Tzafestas and Kapsiotis 1994, Younis and Mahmoud 1986). Systems Modelling deals with general modelling issues of this class of systems, such as how to represent, quantify, and measure - cooperation, coordination, synchronization, and integration (Little 1992, Morris 1967, and Pritsker 1997). Systems Engineering describes methodologies for structuring systems as these are implemented in various application domains (Blanchard and Fabrycky 1990). System Integration deals with achieving common interface within and between different components at various levels of hierarchy in an enterprise (Shaw et al. 1992), as well as different architectures and methodologies (ISO TC 184/SC 5/WG 1 1997, IMTR 99, Hirsch 1995) using distributed artificial intelligence and intelligent agents (Gruber 1995, Stumptner 1997, Wooldridge and Jennings 1995).

At the *reductionist* level, a SC configuration must be based on its local as well as global environmental constraints (Fleischanderl et al. 1998). These constraints are partly imposed as the SC negotiates and compromises to adapt to its cooperative behavior (Jennings 1994). Enterprise modeling as a technique has been used effectively in decomposing complex enterprises, such as a SC. Ontologies are defined to describe unique system descriptions of SC that are relevant to specific application domains (Gruninger 1997). The classic problem for a SC is an inventory management problem requiring coordination of product and information flows through a multi-echelon SC. This class of problem has been solved by integration of the front and back-ends of the SC with costs and lead times as key measures of its performance (Clark 1972, Clark and Scarf 1960, Diks et al. 1996, Diks and Kok 1997, 1998, Hariharan and Zipkin 1995, Pyke and Cohen 1990).

Analytic approach for the general class of supply chains has its origins in economic models of supply and demand coordination. Game Theory principles for payoffs among market competitors have been used effectively to design competitive strategies for supply chains (Gupta and Loulou 1998, Masahiko 1984). Coordination, dealing with interfaces between strategies, objectives and policies for various functions of an enterprise, has received much attention in optimizing the performance of a SC (Malone 1987, Malone and Crowston 1994; Thomas and Griffin 1996, Whang 1995). Various aspects of cooperation have been prescribed for effective management of supply chains (Sousa et al. 1999).

Starting from the evaluation of existing enterprise integration architectures (CIMOSA, GRAI/GIM and PERA), the IFAC/IFIP Task Force on Architectures for Enterprise Integration has developed an overall definition of a generalized architecture framework, GERAM - Generalized Enterprise Reference Architecture and Methodology (ISO TC 184/SC 5/WG 1 1997).

As noted in Section 2, although systems employing mass customization have much in common with general advanced manufacturing systems, issues such as customer involvement call for additional problem-oriented analysis, which is conducted on basis of mass customization oriented reconfiguration methodologies.

4.2   Reconfigurable systems and mass customization

Implementation of mass customization triggers reconfiguration of manufacturing operations and supply network (Eastwood 1996; Kotha 1996). Systematic methodologies for system reconfiguration in the mass customization framework have been investigated by several authors.

Fulkerson (1997) argues that decentralization of control mechanisms in a networked structure such as SC, enables quick response to changing customer requirements. Units can be added, dropped or their functionality adjusted according to current market requirements. The firm is viewed as a web of processes that are executed by autonomous agents. These processes represent the entire range of SC functions including procurement, manufacturing and distribution. An integrated information system supports communications among agents. The author discusses usage of agent technologies and methods of artificial intelligence for product flow management.

Salvador et al. (2002b) develop the road map for mass customization implementation. Turning points of this road map are required variety and volume with the product family, selection of the appropriate product architecture and allocation of component production to suppliers. Svensson and Barfod (2002) describe major steps leading to mass customization. The first step is evaluation of readiness for mass customization. Selection of the mass customization strategy is made at the second step. Organizational capabilities are aligned with customization requirements in Step 3. The final step involves building knowledge needed to implement mass customization. Alfnes and Strandhagen (2000) describe the control model based methodology, which have proven its efficiency in several practical implementations of mass customization. The main principle behind the methodology is decentralization of control mechanisms. Agility, postponement and information availability are three other principles governing this methodology. Efficiency of the system depends upon the ability of semi-autonomous units to cooperate and to adjust to customer requirements.

Integration of transaction tracking and decision support systems to achieve the agility needed for manufacturing of customized products is investigated by Yao and Carlson (2003). The transaction tracking system provides the decision support system with real time data about manufacturing processes. The decision support system uses these data to efficiently adjust manufacturing schedules to current requirements.

The articles surveyed above demonstrate that system reconfiguration is undertaken and mass customization is implemented in response to market conditions and customer requirements. Identification of the need for customization and design of customized products are first steps in any methodology for implementation of customization. Additionally, the articles highlight importance of integrating information management, decision making and knowledge management functions.

However, there is a gap between domain-independent and problem-specific problem-solving approaches. This gap is created by insufficient integration between both approaches (i.e., how exactly domain-independent approaches can be applied in the mass customization framework).

4.3   Motivation and focus of proposed research

The motivation and focus of the research methodology proposed in this Chapter is to integrate the above problem-solving approaches in the design of proposed mass customization methodologies. It is characterized by two main purposes: general and specific.

The *general* purpose is to develop a common body of inter-disciplinary knowledge to understand issues and problems related to reconfigurable systems.

The *specific* purpose is to (a) develop methodology and tools for SC reconfiguration, (b) elaborate framework for knowledge based problem analysis and model building, and (c) quantify factors influencing SC reconfiguration.

The general problems in reconfigurable systems can be classified as related to its market environment, availability of appropriate modeling tools, interconnectedness of decisions at various levels of SC, and availability of common knowledge throughout the system. These can be stated as follows:

- Increasing competitive pressures and consumer focus requires innovative SC modeling and management tools,
- SC modeling tools must capture complex interactions within SC,
- SC configuration decisions have major impact on other decisions at all levels, and
- Knowledge assumes a critical role in a firm's success, and, therefore must be captured, organized and utilized effectively.

Problem solving strategies applied to reconfigurable manufacturing systems entail developing (a) domain independent solution(s) templates at the macro level, (b) capability models for application specific domain dependent problems at micro level, and (c) coordination models to integrate models developed in (a) and (b).

5. INTEGRATED MODELING APPROACH

For improving SC effectiveness, a decision support system that integrates information technology with simulation and optimization techniques is described below. It is built upon a generic methodology proposed by Chandra and Grabis (2002), which emphasizes commonality of general structure of problems across various industry domains. The unique differences that are, however, found in specific problem for a particular industry can be represented by parameters that uniquely set it apart from other problems (in other industry domains).

The decision support system (DSS) consists of two integral parts – information support system (ISS) and decision modeling system (DMS). The DSS is organized to support SC managerial activities and functions to make it streamlined and managed effectively. The DMS is concerned with describing situations in SC domain, building models, analyzing these models and finding solutions for domain problems. The ISS is related to providing adequate information support to DMS. It also provides information about customer requirements and product design. Figure 3-2 depicts the overall picture of DSS components.

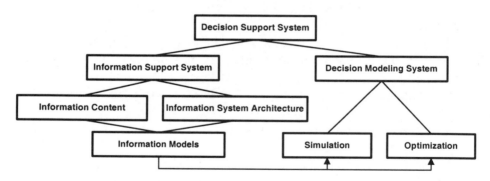

Figure 3-2. Decision support system organization

In order to effectively support DMS, ISS should organize and maintain the information content and provide appropriate infrastructure for delivering information to problem solving simulation and analytical tools.

5.1   Overall framework

The methodology consists of a number of consecutive tasks including definition of the supply chain management problem, generation of alternative SC configurations, decision modeling and final decision making as depicted in Figure 3-3. The main tools utilized in order to accomplish these tasks are

SC process model, modeling database, decision modeling system and SC knowledge library. The modeling methodology is built around a unified representation of the SC modeling problem. This representation is stored in the modeling database. It can be transformed in a number of different formats suitable for accomplishing different modeling tasks. An implementation of the decision support system is developed on the basis of standard specialized software tools. An Enterprise Resource Planning software (SAP/R3, developed by SAP Inc., Germany) provides data storage support. ARIS (developed by IDS-Scheer Inc., Germany) is used in process modeling. MS Excel (developed by Microsoft Inc., USA) is used as a core component of the decision model. It invokes packages such as LINGO (developed by LINDO Systems, USA) and ARENA (developed by Rockwell Software Inc., USA) for solving of specific Linear Programming and Simulation modeling tasks, respectively.

For the manufacturing industry, a supply chain management problem may pertain to supplier selection, demand management, order management, inventory control issues etc. Alternative SC configurations may identify a SC representation that recognizes availability of alternative suppliers, based on criteria such as lower cost, quality, material specifications, etc. A decision model will comprise of a problem representation, problem-solving algorithm(s) and parameters associated with the specific problem under study. A process model for a selected SC configuration will explicitly define processes (or activities) for all business entities such as, manufacturers, component manufacturers, wholesalers/distributors, retailers, and consumers (along with their characteristics captured as attributes and associated variables and parameters), from supply to demand stages. A SC knowledge library will represent the knowledge captured for SC configurations specific to a problem domain. This knowledge can be reused when problems with similar features are encountered in the future.

The unified representation of the SC problem assures data integrity. Modeling results are stored in the SC modeling database and are available to aid solving SC modeling problems with similar properties. Sharing of the modeling database over the Internet provides SC members with access to data. Utilization of generic decision models reduces modeling resource requirements and allows direct comparison of multiple configurations.

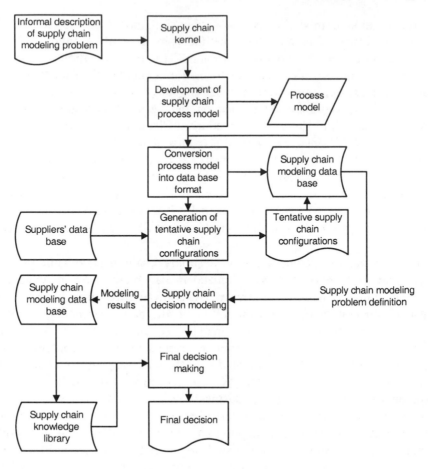

Figure 3-3. The proposed modeling methodology

(Source: Chandra and Grabis, 2002)

5.2   Modeling methodology

The proposed methodology systematically evaluates a specific problem by designing, modeling, and analyzing it according to its context in relation to the total enterprise represented by a SC entity. The step-by-step problem evaluation adopted in this methodology is described below.

Process model and modeling database

An initial state of modeling is either a current SC, which should be reconfigured, or a product or service specification, for which a SC is needed.

The initial state generally is described in an informal manner. A SC kernel provides the template describing data requirements for the formal SC problem definition. It also incorporates information regarding customer requirements and product design. SC models are developed using the kernel data. Development of a SC process model is the next step of SC modeling. The process model provides an exhaustive representation of the general SC management problem. It provides input data for other SC models. Therefore, construction of the process model follows a set of rules elaborated to provide a consistent and generalized representation of the SC (Chandra and Marukyan 2002), which can be transformed in other required data processing formats. The process model describes the initial supply chain management problem referred to as the base configuration. The process model development is accompanied by creation of a SC modeling database. This database is used to store and disseminate the definition of the SC modeling problem and modeling results. The database is created using a template of the standardized SC modeling database model (Chandra and Chilov 2001). The database is created using data extracted from the process model.

Tentative configurations.

The proposed SC modeling methodology is primarily aimed at dealing with a class of problems involving SC reconfiguration. This includes evaluation of alternative SC configurations generated using information from the suppliers' database. Tentative configurations are stored in the SC modeling database, where they are identified as alternative configurations to the base configuration.

Decision modeling

Selected tentative SC configurations are evaluated using the decision modeling system. The decision modeling system establishes the SC configuration from the set of the preliminary selected units by means of optimization and emulates consequences of adopting particular SC configuration and management policies by means of simulation. Each tentative configuration representing the SC modeling problem is retrieved from the decision modeling database. The relational representation of the SC modeling problem is transformed into a format readable by decision model components. This format uses incidence matrices to represent the SC modeling problem. The simulation model is designed using the generic modeling approach (Chandra et al. 2000), which is particularly well suited to deal with modeling a variety of SC configurations. Each configuration is evaluated under a number of different scenarios. A scenario describes different supply chain management policies, environmental and system parameters. Obtained results are stored in the database.

Decision making and knowledge accumulation.

After the decision modeling process is completed with all tentative SC configurations, the modeling database stores a large number of alternative solutions to the initial supply chain management problem. A management problem is to find a decision to be implemented. The problem is exacerbated by the fact that not all factors can be adequately represented in models and several solutions may yield similar performances. This decision making can be assisted by utilization of an enterprise knowledge library.

If the decision modeling process is repeated routinely, then enterprise level SC modeling knowledge is accumulated in the modeling database. This accumulated knowledge is used to create the knowledge library, which stores information about management problems solved, alternative configurations considered, obtained modeling results, implemented decisions and implementation appraisal. The decision implementation appraisal is the crucial step to assure usability of accumulated knowledge. New decisions can be matched to stored decisions and those, which are likely, to yield unsatisfactory implementation results, can be filtered out using the appraisal criteria. The knowledge library is built utilizing the concept of SC ontology described in Chandra and Tumanyan (2002).

5.3   Information support system

From information organization perspective, SC is viewed as a set of tasks necessary to provide products satisfying customer needs. Each task is associated with some knowledge required for accomplishing these tasks. The information support system is used to capture and formalize this knowledge.

The conceptual framework for information content development is depicted in Figure 3-4. Information conceptualization starts with systematic representation of system existence. It aims to provide a multidisciplinary representation of SC activities and characteristics. The necessity of information standardization and unification is considered as the vehicle of information integration in SC, where the information system is a diverse and heterogeneous environment. The SC information structuring mechanism proposed in the form of the system taxonomy, serves two main objectives:

1. The system taxonomy provides standardization of terms and definition, thus ensuring shared vocabulary across SC domain users and members.
2. The system taxonomy also provides unified structure for a formal representation, thus ensuring that data and knowledge can be represented in a format consumable by SC members' software applications.

SC operational specific is problem-orientation. Therefore, activities and their information models are to be identified to support these activities. SC activities can be decomposed into problems and tasks. Decision modeling applications are designed for solving these problems. If ISS delivers to DMS problem object models, those can effectively process information and knowledge encapsulated into these objects and generate solutions for solving problems. Each problem is to be investigated in terms of finding parameters involved in it. Once these are found, the system taxonomy can be scanned to find them. A projection from the system taxonomy will reflect only identified parameters, thus generating the problem model. Projecting various problems from the same source provides consistency and compatibility of various informational representations and gives DMS the standard specification to work with.

Development of the problem taxonomy results in an abstract hierarchy of characteristics. But DMS needs variables with values to operate with. Ontology is proposed as a tool for information / knowledge capture, assembly, store, dissemination, and representation. By studying problems for which information models are required to be built, sources of data, rules and regulations can be identified. Ontology development is a process, where implicit knowledge captured as a result of a collaborative work of domain experts, knowledge engineers, and software programmers, are encoded in a common programming language, thus turning implicit into explicit knowledge.

Another layer that still exists in the information modeling framework is the object model, which is an instance of ontology. Each time a DMS software makes a request of a particular ontology, it receives the ontologies object model, which can be different from the request made at another point of time, since ontology is dynamic construct, where data are changed every instance, and rules can be changed, too. The object model is a tangible software construct, which can be directly embedded into DMS applications.

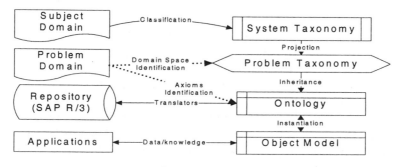

Figure 3-4. Information content modeling

5.4   Decision modeling system

DMS addresses some of the deficiencies, such as limited scope, treatment of uncertainty and SC dynamics of existing configuration models (see Ballou (2001) for summary on challenges in SC configuration). That is achieved by combining strategic optimization and operational level simulation models (Figure 3-5).

The optimization model is used to establish the SC configuration. The simulation model is used to evaluate the obtained configuration at the operational level. This approach allows for assessing the configuration decisions subject to stochastic and dynamic factors. Specifically, time related factors, which are essential in mass customization and not supported by conventional configuration models, can be analyzed.

Both optimization and simulation models retrieve the data necessary for modeling from the information support system. Ontological translators are used to arrange data in formats suitable for decision modeling purposes. Modeling results are sent back to the information support system in order to convert these into SC modeling knowledge.

The standard SC configuration model optimizes cost of establishing and operating the SC subject to demand satisfaction, production balance and resource utilization constraints. It determines which units are to be included in the SC, their location and size, and establishes product flows among units. The optimization models can be customized to include problem specific constraints and decision variables. Regardless of the possible modifications, the optimization model has requirements on data it should pass to the simulation model. These data are: 1) units included in the SC, 2) links between SC units, 3) production volume for a product at a particular unit, and 4) capacity for each resource.

The simulation model is automatically generated according to optimization results on the basis of a predefined template model. The model is used to simulate performance of the obtained SC configuration at the operational level. The performance is evaluated according to various criteria such as cost, delivery time and customer service, and under multiple scenarios in order to evaluate flexibility and robustness of the system. Simulation is performed using short-term customer demand series, which are generated according to specified demand models. During the simulation process, it interacts with inventory / production planning and forecasting models. The forecasting model provides short-term forecasts of the generated external demand (one can use actual demand series, if forecasting errors are not considered as the major factor). The inventory / production planning model is used in determining size and timing of inventory replenishment orders according to demand forecasts. Currently, inventory / production planning is performed using a materials requirements planning (MRP) based model. Analytical operational planning models are also aimed to be replaceable to represent custom user requirements.

The simulation model has a relatively high level of abstraction mainly associated with reduction of modeling efforts and information availability. Additionally, it is reasonable to expect that such level of abstraction is sufficient during strategic decision making.

Detailed descriptions of the optimization model and the simulation model are given in the following sub-sections.

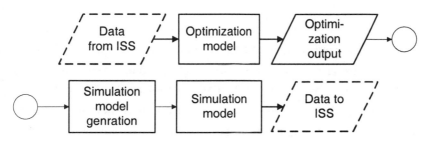

Figure 3-5. The SC configuration decision model

Optimization model

The standard SC configuration model optimizes cost of establishing and operating the SC subject to demand satisfaction, production balance and resource utilization constraints. It determines which units are to be included in the SC, their location and size, and establishes product flows among units. A general SC configuration optimization models is given as

$$TC = \sum_i^N \sum_j^M (c_{ij}^1 + c_{ij}^2) * X_{ij} + \sum_i^N \sum_j^M \sum_{j'}^M c_{ijj'}^3 Q_{ijj'}$$
$$+ \sum_j^M c_j^4 H_j + \sum_j^M c_j^5 W_j \to \min$$

subject to

$$\sum_j^M X_{ij} - \sum_i^N \sum_j^M \sum_{j'}^M Q_{ijj'} = D_i, \ i = 1,...,N$$

$$\sum_i^N \sum_j^M \delta_{ii'} Q_{ijj'} = X_{i'j'}, \ i' = 1,...,N, \ j' = 1,...,M$$

$$X_{ij} < \gamma_{ij}, \ i = 1,...,N, \ j = 1,...,M$$

$$\sum_{i=1}^{M} X_{ij} < H_j, j = 1, \ldots, M$$

$$\sum_{i=1}^{N} X_{ij} < PW_j, j = 1, \ldots, M$$

$$W_i \in \{0,1\},$$

where M is number of SC units and N is number of products (in the case of mass customization, product families are used instead of products). X_{ij} is quantity of the ith product produced by the jth unit, $Q_{ijj'}$ is quantity of the ith product shipped from the jth unit to the j'th unit. c_{ij}^1, c_{ij}^2, $c_{ijj'}^3$, c_j^4 and c_j^5 are processing, purchasing, transportation (for product between two units), capacity cost and fixed supply link maintenance costs, respectively. H_j is capacity of the jth unit. W_j indicates whether a unit is included in the SC, D_i is external demand for the ith product. $\delta_{ii'}$ defines the number of items of the ith product needed to assemble the i'th product. γ_{ij} is the production capacity for the ith product at the jth unit. P is a large constant number. X_{ij}, $Q_{ijj'}$, H_j and W_i are decision variables.

The optimization models can be modified to include problem specific constraints and decision variables. The most commonly considered modifications relate to treatment of capacity and distinguishing between types of SC units (e.g., suppliers, plants, distribution centers). Regardless of possible modifications, the optimization model passes to the simulation model, following data: 1) units included in the SC, 2) links between SC units, 3) production volume for a product at a particular unit, and 4) capacity for each resource.

Simulation model

The simulation model is used to simulate performance of the obtained SC configuration at the operational level. The performance is evaluated according to various criteria such as cost, delivery time and customer service, and under multiple scenarios in order to evaluate flexibility and robustness of the system. Simulation is performed using short-term customer demand series, which are generated according to specified demand models. During the simulation process, it interacts with analytical operational inventory / production planning and forecasting models. The forecasting model provides short-term forecasts of the generated external demand (in the case of mass customization, actual demand or forecasts for base products are used depending upon a particular mass customization strategy). The inventory / production planning model is used to determine size and timing of inventory replenishment orders.

The simulation model is automatically generated according to optimization results. Besides the output data from the optimization model, the simulation model also requires simulation specific data, which describe operational level characteristics of SC units such as processing time, transportation time, inventory holding and production setup costs.

The simulation model is developed using a specially designed simulation model building approach. This approach is developed in order to reduce model building efforts. It assumes that SC units at a certain level of abstraction have common functionality. These common functions are handling of incoming and outgoing flows, flow transformation and control. Such generic representation of a SC unit implemented in ARENA is shown in Figure 3-6. Unique characteristics of SC units are represented through parameters and control mechanisms used. The SC is constructed as a network of generic units.

The simulation model has two control levels: the global or chain level and the local or unit level. Some of the control functions are implemented using standard ARENA tools, while others are implemented as Visual Basic procedures integrated in the model using the VBA block. The simulation process at the chain level is initiated by generating an entity at the beginning of each micro period. This entity triggers elaboration or updating of the inventory / production schedule.

The SC base configuration is automatically generated on the basis of a predefined model template. The template does not contain any objects. However, it contains the standardized code for executing both local and global control functions. The template is an ARENA .mod file supplemented with Visual Basic code. The template is populated with simulation objects using a custom-built Visual Basic program through the ActiveX automation. Objects are generated according to optimization results and data provided by ISS. These data are organized using Microsoft Excel spreadsheets. The generated model only embeds the core of SC structure – product to unit assignments. Other structural data, such as bill of materials, resource to unit assignments and processing times are read from input data files on initializing a simulation run. These data can also be changed during the simulation run. That allows implementing relatively large changes in the simulation model only by modifying input data but not the model structure.

Each product is held in its own queue, and the holding condition is also product specific. These queues and holding conditions are organized using the set of queues, and the set of expressions option, respectively. Such sets are

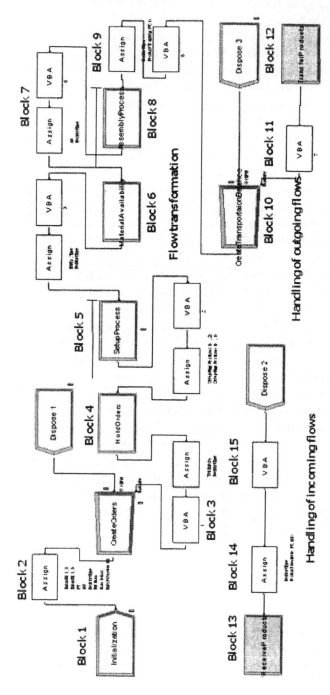

Figure 3-6. Representation of the generic SC unit (ARENA implementation)

useful for handling a large variety of products as in the case of mass customization.

The generated model is the conventional ARENA model. A user can edit the model, use the standard output reporting features and perform other manipulations.

At the beginning of simulation, modeling data from the data model are loaded in the simulation model. Some of the data tables are loaded into ARENA arrays for access by ARENA objects, while some others are loaded in Visual Basic arrays for access by control functions. All VBA blocks invoke a main Visual Basic procedure. An entity attribute characterizing the function to be performed is assigned to the entity before it enters a VBA block. The main procedure reads attributes of the entity to determine a specific procedure to be called and parameters of the specific procedure. The specific procedures read data from the production schedule, reassign resources, update inventory data, check material availability, determine transportation route for products, etc. All procedures are part of the model template. The model generator is implemented using Visual Basic.

Analysis

DMS has been applied in solving various SC configuration problems. Relationships between supplier selection, which is the strategic decision, and inventory management and short-term forecasting, which are operational decisions, have been of particular interest (Chandra and Grabis 2002, 2003).

The strategic level optimization models has been used to select suppliers, while the operational level simulation model, which also interacts with forecasting and inventory management models, has been used to evaluate the supplier selection with regard to operational factors. Impact of the supply lead time, which is difficult to evaluate by mathematical programming models, on SC performance is of particular interest because this lead time influences the accuracy of demand information, which can be used in planning materials procurement.

Figure 3-7 summarizes findings about interactions among decisions. Supplier selection depends upon external links from product and process design processes and from customer demand forecasting. Product and process design processes determine which materials (modules) need to be procured. The long-term forecasts determine quantity of these materials required (even though end-products are assembled upon receiving customer orders, sufficient supply base capacity should be reserved). Supplier selection influences further manufacturer's inventory management decisions through the supply lead time parameter. Therefore, the strategic level decision making also accounts for the operational factors. The influence of supplier selection on inventory management depends upon the inventory management policy chosen, which in turn determines the choice of the forecasting method. For instance, MRP based policies are better suited for handling time-varying demand with predictable pattern than re-order point policies. However, full potential of these methods can only be achieved, if appropriate forecasting methods adequately characterizing the demand pattern are used. Inventory management for the manufacturer also depends upon distribution requirements. After the inventory management costs for the particular supplier selection have been evaluated using operational models, expected inventory management costs are fed back to the optimization model for more accurate assessment of suppliers.

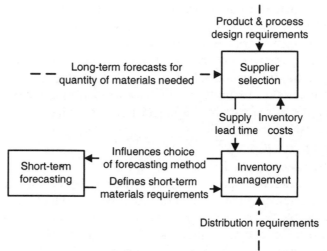

Figure 3-7. Interactions among selected SC management decisions

6. CONCLUSION

Interrelatedness between mass customization and the visionary advanced manufacturing system is demonstrated. The compatibility of both concepts indicates that mass customization is the natural way of development in manufacturing. At the same time, mass customization is also not an exclusive alternative to mass production and there are opportunities for both approaches and mutual benefit.

Methodologies for implementing mass customization are analyzed from the system reconfiguration perspective, which is central to the advanced manufacturing system. The decision support system integrating information and decision modeling components is proposed to address crucial issues in SC reconfiguration. The information support system maintains system integrity and allows for generation of modeling knowledge, which describes interactions among various problem areas including strategic planning, forecasting and inventory control and among SC members. The decision modeling system provides tools for validation of SC configuration decisions with respect to demand uncertainty and system responsiveness, which are essential factors for evaluation of systems adopting mass customization.

REFERENCES

1. Aldanondo M., Hadj-Hamou K., Moynard G., Lamothe J. Mass customization and configuration: Requirement analysis and constraint based modeling propositions. Integrated Computer-Aided Engineering 2003; 10: 177.
2. Alfnes E., Strandhagen J. O. Enterprise Design for Mass Customisation: The Control Model Methodology. International Journal of Logistics: Research and Applications 2000; 3: 111-125.
3. Alford D., Sackett P., Nelder G. Mass customisation -- an automotive perspective. International Journal of Production Economics 2000; 65: 99-110.
4. Ballou R. H. Unresolved issues in supply chain network design. Information Systems Frontiers 2001; 3: 417-426.
5. Blanchard S. B., Fabrycky W. J., *Systems Engineering and Analysis*. New Jersey: Prentice Hall, 1990.
6. Chandra C. Enterprise architectural framework for supply-chain integration. Proceedings of the 6th Annual Industrial Engineering Research Conference, May 17-18, Miami Beach, Florida, 1997; 873-878.
7. Chandra C., Chilov N. Object-Oriented Data Model for Supply Chain Configuration Management. Proceedings: Tenth Annual Industrial Engineering Research Conference (IERC), May 20-22, Dallas, Texas, 2001.
8. Chandra C., Grabis J. Modeling Floating Supply Chains. Proceedings: Eleventh Annual Industrial Engineering Research Conference (IERC), May 19-22, Orlando, Florida, 2002.
9. Chandra C., Grabis J. Impact of supply lead time on supplier selection: An operational perspective. Proceedings: Twelfth Annual Industrial Engineering Research Conference (IERC), May 18-20, Portland, Oregon, 2003.
10. Chandra C., Marukyan R. Elaborating Process Models for Supply Chain Reconfiguration. Proceedings: Eleventh Annual Industrial Engineering Research Conference (IERC), May 19-22, Orlando, Florida, 2002.
11. Chandra C., Nastasi A. J., Norris T. L., Tag P. Enterprise Modeling for Capacity Management in Supply Chain Simulation. Proceedings: Ninth Industrial Engineering Research Conference (IERC), May 21 - 24, Cleveland, Ohio, 2000.
12. Chandra C., Tumanyan A. Supply Chain Reconfiguration: Designing Information Support with System Taxonomy Principles. Proceedings: Eleventh Annual Industrial Engineering Research Conference (IERC), May 19-22, Orlando, Florida, 2002.
13. Clark A. J. An informal survey of multi-echelon inventory theory. Naval Research Logistics Quarterly 1972; 19: 621-650.
14. Clark A. J., Scarf H. Optimal policies for multi-echelon inventory problem. Management Science 1960; 6: 475-490.
15. Cohen M. A., Lee H. L. Strategic Analysis of Integrated Production-Distribution System. Operations Research 1988; 36: 216.
16. Cohen M. A., Lee H. L. Resource deployment analysis of global manufacturing and distribution networks. Journal of Manufacturing Operations Management 1989; 2: 81-104.
17. Contractor F. J., Lorange P. The growth of alliances in knowledge-based economy. International Business Review 2002; 11: 485-502.

18. Da Silveira G., Borenstein D., Fogliatto F. Mass customization: Literature review and research directions. International Journal of Production Economics 2001; 72: 1-13.

19. Deleersnyder J. L., Hodgson T. J., King R. E., O'Grady P. J., Savva A. Integrating kanban type pull systems and MRP type push systems: insights from a Markovian model. IIE Transactions 1992; 24: 43-56.

20. Diks E. B., De Kok A. G. Optimal control of a divergent multi-echelon inventory system. European Journal of Operational Research 1997.

21. Diks E. B., De Kok A. G. Computational results for the control of a divergent N-echelon inventory system. International Journal of production economics 1998.

22. Diks E. B., De Kok A. G., Lagodimos A. G. Multi-echelon systems: a service measure perspective. European Journal of Operational Research 1996; 95: 241-263.

23. Drew S. A. A. The application of hierarchical control methods to a managerial problem. International Journal of Systems Science 1975; 6: 371-395.

24. Eastwood M. A. Implementing mass customization. Computers in Industry 1996; 30: 171-174.

25. Fleischanderl G., G. F., A. H., H. S., M. S. Configuring Large Systems Using Generative Constraint Satisfaction. IEEE Intelligent Systems and their application 1998; 1: 59-68.

26. Fowler S. W., King A. W., Marsh S. J., Victor B. Beyond products: new strategic imperatives for developing competencies in dynamic environments. Journal of Engineering and Technology Management 2000; 17: 357-377.

27. Fujita K. Product variety optimization under modular architecture. Computer-Aided Design 2002; 34: 953-965.

28. Fulkerson F. A response to dynamic change in the market place. Decision Support Systems 1997; 21: 199-214.

29. Furst K., Schmidt T. Turbulent markets need flexible supply chain communication. Production Planning & Control 2001; 12: 525-533.

30. Graves S. C. Using Lagrangean techniques to solve hierarchical production planning problems. Management Science 1982; 28: 260-275.

31. Griffiths J., Margetts D. Variation in production schedules -- implications for both the company and its suppliers. Journal of Materials Processing Technology 2000; 103: 155-159.

32. Gruber T. Toward principles for the Design of Ontologies Used for Knowledge Sharing. International Journal of Human and Computer Studies 1995; 43: 907-928.

33. Gruninger M., *Integrated Ontologies for Enterprise Modelling*. In *Enterprise Engineering and Integration. Building International Consensus*, K.Kosanke and J.Nel, ed. Springer, 1997; 368-377.

34. Gu X. J., Qi G. N., Yang Z. X., Zheng G. J. Research of the optimization methods for mass customization (MC). Journal of Materials Processing Technology 2002; 129: 507-512.

35. Gunasekaran A., Goyal S. K., Martikainen T., Yli-Olli P. A conceptual framework for the implementation of zero inventory and just in time manufacturing concepts. Human Factors and Ergonomics in Manufacturing 1998; 8: 63-78.

36. Gupta S., Loulou R. Process Innovation, Product Differentiation, and Channel Structure: Strategic Incentives in a Duopoly. Marketing Science 1998; 17.

37. Hackman S. T., Leachman R. C. A general framework for modeling production. Management Science 1989; 35: 478-495.

38. Hariharan R., Zipkin P. Customer-order information, leadtimes, and inventories. Management Science 1995; 41: 1599-1607.

39. He Q.-M., Jewkes E. M. Performance measures of a make-to-order inventory-production system. IIE Transactions 2000; 32: 409-419.
40. Hirsch B. Information System Concept for the Management of Distributed Production. Computers in Industry 1995; 26: 229-241.
41. Hirschhorn L., Noble P., Rankin T. Sociotechnical systems in an age of mass customization. Journal of Engineering and Technology Management 2001; 18:241-252.
42. IMTR. Technologies for Enterprise Integration, Rev 3.1. Integrated Manufacturing Technology Roadmapping Project, Oak Ridge Centers for Manufacturing Technology, Oak Ridge, Tennessee.
43. ISO TC 184/SC 5/WG 1. Requirements for enterprise reference architectures and methodologies. http: //www.mel.nist.gov/sc5wg1/gera-std/ger-anxs.html, 1997.
44. Jennings R., Cooperation in Industrial Multi-agent Systems. World Scientific Series in Computer Science 43. World Scientific Publishing Co., 1994.
45. Jiao J., Tseng M. M. Customizability analysis in design for mass customization. Computer-Aided Design, 2004, In press.
46. Joneja A., Lee N. A modular, parametric virbratory feeder: A case study for flexible assembly tools for mass customization. IIE Transactions 1998; 30: 923-931.
47. Karsak E. E., Kuzgunkaya O. A fuzzy multiple objective programming approach for the selection of flexible manufacturing system. International Journal of Production Economics 2002; 79: 101-111.
48. Kim J. Hierarchical structure of intranet functions and their relative importance: using the analytic hierarchy process for virtual organization. Decision Support Systems 1998; 23: 59-74.
49. Klir G. J., Facets of Systems Science. New York: Plenum Press, 1991.
50. Kotha S. From Mass Production to Mass Customization: The Case of the National Industrial Bicycle Company of Japan. European Management Journal 1996; 14: 442-450.
51. Lee H. L. Effective inventory and service management through product and process redesign. Operations Research 1996; 44: 151-159.
52. Lee H. L., Billington C. Material management in decentralized supply chains. Operations Research 1993; 41: 835-847.
53. Little J. D. C. Tautologies, Models and Theories: Can We Find "Laws" of Manufacturing? IIE Transactions 1992; 24: 7-13.
54. Ma S., Wang W., Liu L. Commonality and postponement in multistage assembly systems. European Journal of Operational Research 2002; 142: 523-538.
55. Malone T. W. Modeling coordination in organizations and markets. Management Science 1987; 33: 1317-1332.
56. Malone T. W., Crowston K. The interdisciplinary study of coordination. ACM Computing Surveys 1994; 26: 87-119.
57. Masahiko A., The Co-operative Game Theory of the Firm. Oxford: Oxford University Press, 1984.
58. Morris W. T. On the art of modeling. Management Science 1967; 13: B707-B717.
59. NIST. Manufacturing Enterprise Integration Program. National Institute of Standards and Technology: Gaithersburg, 1999.
60. National Research Council. Visionary Manufacturing Challenges for 2020, Committee on Visionary Manufacturing Challenges, National Academy Press: Washington D.C., 1998.
61. Pritsker A. A. B. Modeling in performance-enhancing processes. Operations Research 1997; 45: 797-804.
62. Pyke D. F., Cohen M. A. Push and pull in manufacturing and distribution systems. Journal of Operations Management 1990; 9: 24-43.

63. Roy U., Kodkani S. S. Product modeling within the framework of the World Wide Web. IIE Transactions (Institute of Industrial Engineers) 1999; 31:667-677.

64. Salvador F., Forza C., Rungtusanatham M. Modularity, product variety, production volume, and component sourcing: theorizing beyond generic prescriptions. Journal of Operations Management 2002a; 20: 549-575.

65. Salvador F., Forza C., Rungtusanatham M. How to mass customize: Product architectures, sourcing configurations. Business Horizons 2002b; 45: 61-69.

66. Shaw M. J., Solberg J. J., Woo T. C. System integration in intelligent manufacturing: an introduction. IIE Transactions 1992; 24: 2-6.

67. Sousa P., Heikkila T., Kollingbaum M., Valckenaers P. Aspects of co-operation in Distributed Manufacturing Systems. Proceedings of the Second International Workshop on Intelligent Manufacturing Systems, 1999; 685-717.

68. Stumptner M. An overview of knowledge-based configuration. AI Communications 1997; 10: 111-125.

69. Svensson C., Barfod A. Limits and opportunities in mass customization for "build to order" SMEs. Computers and Industry 2002; 49: 77-89.

70. Swaminathan J. M., Smith S. F., Sadeh N. M. Modeling supply chain dynamics: A multiagent approach. Decision Sciences 1998; 29: 607.

71. Thomas D. J., Griffin P. M. Coordinated supply chain management. European Journal of Operational Research 1996; 94: 1-15.

72. Tseng M. M., Jiao J. Case-based evolutionary design for mass customization. Computers and Industrial Engineering 1997; 33: 319-323.

73. Tseng M. M., Jiao J., Su C. J. Virtual prototyping for customized product development. Integrated Manufacturing Systems 1998; 9: 334-343.

74. Turowski K. Agent-based e-commerce in case of mass customization. International Journal of Production Economics 2002; 75: 69-81.

75. Twede D., Clarke R. H., Tait J. A. Packaging postponement: A global packaging strategy. Packaging Technology and Science 2000; 13: 105-115.

76. Tzafestas S., Kapsiotis G. Coordinated control of manufacturing/supply chains using multi-level techniques. Computer Integrated Manufacturing Systems 1994; 7: 206-212.

77. Weng K. K. Strategies for integrating lead time and customer-order decisions. IIE Transactions 1999; 31: 161-171.

78. Whang S. Coordination in operations: a taxonomy. Journal of Operations Management 1995; 12: 413-422.

79. Wooldridge M., Jennings N. R. Intelligent agents: theory and practice. The Knowledge Engineering Review 1995; 10: 115-152.

80. Worren N., Moore K., Cardona P. Modularity, strategic flexibility, and firm performance: A study of the home appliance industry. Strategic Management Journal, 2004, In press:

81. Yao A. C., Carlson J. G. H. Agility and mixed-model furniture production. International Journal of Production Economics 2003; 81-82: 95-102.

82. Younis M. A., Mahmoud M. S. Optimal inventory for unpredicted production capacity and raw material supply. Large Scale Systems 1986; 11: 1-17.

CHAPTER 4

LOGISTICS AND SUPPLY CHAIN MANAGEMENT FOR MASS CUSTOMIZATION

Charu Chandra[1], Jānis Grabis[2]

[1]*University of Michigan-Dearborn*
[2]*Riga Technical University*

Abstract: Efficient logistics and supply chain management is one of the key preconditions for adopting mass customization strategies. However, not much published research exists on analysis and integration of logistics and supply chain management activities in the mass customization framework. This Chapter surveys the existing literature to analyze interrelationships between supply chain management and mass customization strategies. Models for supply chain configuration and inventory management addressing some of the key issues in logistics management for mass customization is discussed.

Key words: Logistics, supply chain management, supply chain configuration, third-party logistics, postponement, information technology.

1. INTRODUCTION

The success in implementing mass customization strategies depends upon multiple interrelated mass customization enablers identified in Chapter 1. These enablers include agile manufacturing, customer driven design and manufacturing, advanced manufacturing technologies and information technology (Da Silveria et al. 2001). Similarly, logistics and supply chain management are also two of the most important enablers because they enable minimization of product cost, increasing product variety, and shortening time-to-market.

According to Mentzer et al. (2001), key functional areas of supply chain management are marketing, sales, research and development, forecasting, production, purchasing, logistics, information systems, finance and customer service. The key feature of supply chain management is dealing with the cross-organizational character of supply chain. It can be observed that logistics is one of the functional areas of supply chain management because it provides interface with many of the above functions. A broader definition of logistics includes functional areas, such as customer service, demand forecasting, inventory control, distribution communications, facility location, procurement and transportation (Lambert et al. 1997). Some of these functional areas (e.g., inventory control, transportation) are sub-problems of the logistics area of supply chain management, while others (e.g., forecasting, purchasing) overlap with functional areas of supply chain management. (See Lambert and Cooper (2000) for a detailed discussion on difference between logistics and supply chain management.) This Chapter perceives logistics as an integral part of supply chain management and focuses on those functional areas, which are also included in the broader definition of logistics.

Firms adopting mass customization strategies aim to supply the mass market with products customized according to requirements of individual customers and achieving the efficiency characteristic of mass production. Achieving these conflicting objectives requires joint well-coordinated involvement of all mass customization enablers. Logistics has been referred to as the glue binding other activities together (Gooley 1998). Despite this apparent importance of logistics and supply chain management in mass customization, little research has been done on integrated modeling of all logistics activities in the mass customization framework.

The main objectives of this article are to (a) identify specific issues of logistics and supply chain management that have profound impact on mass customization strategies, (b) survey published research relating supply chain management problems and mass customization, and (c) discuss several models addressing important supply chain management issues in the mass customization framework.

The rest of the Chapter is organized as follows. The following section identifies problem areas of logistics and supply chain management, which are most important for successful adoption of mass customization strategies. Section 3 surveys problem-solving approaches for particular functional areas. Section 4 describes specific supply chain configuration and inventory management models, which address some of the critical issues of supply chain management in the mass customization framework. Future research directions are summarized in Section 5.

2. LOGISTICS ISSUES IN MASS CUSTOMIZATION

Mass customization aims to satisfy needs of individual customer with mass production efficiency. Besides requiring the right product, a customer also requires the right price, acceptable delivery times, quality and service. This Chapter only studies the impact of logistics to providing the right price and the acceptable delivery time. Although logistics and supply chain management also influence meeting of other requirements, other mass customization enablers are assumed to have larger impact on these requirements. Providing the right price can only be achieved by optimizing all logistical activities. By means of the following discussion, it is aimed to identify issues requiring particular attention in managing logistics, while implementing mass customization strategies.

A significant portion of literature surveyed in Chapter 1 of this book has identified supply chain management as one of the enablers of mass customization.

2.1 Logistics Framework

The driving force behind supply chain activities is customer demand transmitted upstream from the point of demand to suppliers. As a response to the external demand, materials and products are moved downstream the supply chain. Logistics and supply chain management are responsible for receiving customer orders and transforming these into production orders. These two activities need to be executed with a little time delay to provide clear understanding of customer requirements. Products ordered are supplied using a distribution network with a high degree of responsiveness and flexibility (short lead times, small lots and low cost). These products are produced at manufacturing facilities from the set of materials procured from selected suppliers. Product design, process design and capacity planning should be aligned to ensure the quick manufacturing process. All types of inventories (materials, work-in-process, end-product) are to be minimized.

Flexibility and lean manufacturing practices play important role in achieving these objectives. The suppliers should provide materials of high quality and on schedule to minimize manufacturing delays. Coordination of activities is required for the supply chain to be efficient, which keeps the supply chain functional. Mass customization has impact on the way these activities are handled. However, this impact varies among activities. Some of them can be carried out in a fashion similar to the case of mass production, while others require particular attention.

2.2   Identification of Issues in Mass Customization

Ahlstrom and Westbrook (1999) conduct a survey of companies in order to identify issues surrounding mass customization with emphasis on operations management. One of the main methods used to meet customer demand for customized products is increasing the range of stock. Such approach indicates that companies try to achieve customization by not changing their production paradigm. Order response time and product delivery time are among the main performance measures used to evaluate impact of mass customization. Weaker supplier delivery performance is among the negative implications of customization. Supply chain management is the second most and often-mentioned difficulty associated with customization. Relatively few companies indicate that information technology hinders the customization process. However, information technology is mentioned as one of the major hurdles to increase the level of customization in future. Probably, that is due to the necessity to expand the information technology support beyond the borders of a single enterprise. The survey also shows that the supply from stock has reached its limits to provide customization. Interestingly, suppliers and distributors are mentioned only as marginal barriers to increasing customization.

Slats et al. (1995) reproduce a classification from Baarends and Witbreuk (1994), showing complexity of logistics depending upon product variety and product value. Large variety of low value products requires special attention to transportation. Large variety of high value products requires maximizing throughput and reducing inventory. By associating large product variety and mass customization, it can be concluded that reducing inventories and optimizing transportation are among the main objectives in managing logistics activities in the mass customization framework.

Silver et al. (1998) have provided an interesting opportunity for evaluation of relationships between mass customization and logistics. Authors in their analysis of manufacturing systems indicate that a job shop approach is the most appropriate for production of customized products. The job shop is characterized by difficult prediction of material requirements, small inventories of raw materials, high inventories of work-in-process, very low inventories of finished products, low control over suppliers and high requirements towards information technology.

Gooley (1998) in his description of relationships between logistics and mass customization indicates that the main emphasis is on minimizing inventory, customer responsiveness and relationships with suppliers. Third-party logistics is advocated as an important factor enabling mass customization by providing flexibility and proximity to the final customer without incurring fixed costs.

Kotha (1996) discusses the classical mass customization example at a bicycle manufacturer. Mass customization has been enabled by having closely located suppliers, which provide frequent deliveries, and by an information system linking suppliers, the manufacturer and retailers. Randall and Ulrich (2001) using empirical data from the U.S. bicycle industry have analyzed relationships between product variety, supply chain structure and performance of companies. They establish an association between the product variety and the supply chain structure. Specifically, it has been shown that, if achieving variety requires large investments in manufacturing equipment, manufacturing is concentrated in a few scale-efficient overseas facilities. However, if variety causes larger mediation costs (associated with higher demand uncertainty) than manufacturing costs, local manufacturing facilities are preferred regardless of scale-efficiency.

Motorola has successfully adopted mass customization strategies (Eastwood 1996). Consolidation of the sub-assembly and the final assembly at one location to save on manufacturing lead-time was the primary step to launch the Bandit pager customization program. Supply chain modeling was used to reduce supply chain uncertainties that caused high inventory levels.

The automotive industry is the largest provider of customized products directly to customers even though the degree of customization varies greatly. Alford et al. (2000) argue that optional customization is the most appropriate type of customization for the automotive industry. Relationships with suppliers characterized by emergence of system suppliers are put forward as an important aspect for enhancing customization capabilities. Authors indicate that many suppliers tend to work in close proximity to OEMs and just-in-time deliveries are frequently used. Fisher and Ittner (1999) describe the automotive assembly process, including installation of custom options. Improved manufacturing technology that allows for shorter setup times and manufacturing flexibility are mentioned as main operational enablers of customization. Just-in-time deliveries are argued as not only reducing inventory, but also speeding up the assembly process by delivering components in the right assembly order.

Feitzinger and Lee (1997) discuss the importance of organizing a supply chain in a manner that enables customer responsiveness and cost competitive operations. They consider postponement as a primary method for delivering customized products. Verwoerd (1999) has provided a conceptual analysis on location of a decoupling point. Customization possibilities at various stages of the supply chain, inventory considerations, speed of production processes, distribution processes, and information processing are mentioned among factors influencing location of the decoupling point. Van Hoek (2000) discusses the impact of postponement on mass customization. The author indicates that completion of product manufacturing is often moved down the distribution channel, which provides an opportunity for third-party logistics providers to expand their services. Tyan et al. (2003) indicate that third-party logistics provide an opportunity to reduce inventory and improve responsiveness by optimizing transportation operations. Rabinovich et al. (2003) show by means of structural equations modeling that customization improves inventory performance. However, the inventory performance is associated only with end-product inventory, not materials inventory.

Information technology has a major role on the success of mass customization (Erens and Hegge 1994). It allows effective coordination between suppliers, manufacturers and customers. Product delivery lead times are not only reduced by faster processing of orders but also by eliminating rework caused by communication errors. Furst and Schmidt (2001) develop Internet based solutions for communicating in the supply chain environment to support mass customization. These solutions are aimed to reduced turbulences in the supply chain. Ghiassia and Spera (2003) identify requirements for information technologies to support mass customization and develop a software system satisfying these requirements. The requirements identified focus on improving coordination among supply chain members.

Resolving many issues in mass customization arises directly from product design. Indeed, majority of product processing decisions including logistics operations are already embedded into the product design (e.g. Jiao et al. 2003, Nevins and Whitney 1989). Dowlatshahi (1999) develops a clustering based methodology for incorporating logistics consideration into the product design.

2. 3 Problems in Logistics and Supply Chain Management for Mass Customization

The main supply chain management and logistics problems to be addressed in order to successfully adopt mass customization strategies are:
1. Facility location and supply chain configuration – frequent deliveries in small lots require a feasible solution for transportation costs

throughout the supply chain. Concentration of facilities is one of the possible solutions. Configuration decisions also include strategic level capacity planning. Capacity planning should be flexible enough to accommodate changes in market conditions.

2. Supplier selection – manufacturers utilizing mass customization policies are highly dependent upon their suppliers regarding delivery promptness, quantities, quality, etc. Therefore, comprehensive supplier selection models are to be used.

3. Postponement – moving the final assembly of the product down the supply chain is one of the frequently used strategies to reduce impact of customer demand uncertainty and to improve responsiveness.

4. Inventory management – inventory is one of the major sources for cost increase due to adoption of mass customization. Short lead times, small batches and uncertainty reduction strategies are used to improve efficiency of inventory control which can be achieved by employing advanced inventory management approaches, such as strategic alliance and third-party logistics providers. Just-in-time deliveries are important. Additionally, coordination plays important role in the multi-tiered supply chain environment, where solutions for information sharing and bullwhip effect issues are essential.

5. Third-party logistics services are employed to improve efficiency of distribution and inventory management as well as save on scarce capital and human resources on managing a warehouse facility.

6. Information technology contributes by eliminating delays associated with processing of orders and managing product information. Additionally, it provides efficient means for exchanging product requirements between customer and manufacturer and between manufacturer and suppliers. Obviously, information technology contributes to solving many other issues of mass customization. The role of information technology beyond logistics and supply chain management is discussed in Chapters 9 of this book.

Figure 4-1 depicts relationships between key issues identified and summarized above. It indicates their relevance to the main logistics and supply chain management areas. The figure does not include coordination, which, essentially, relates to any of the above-mentioned problems.

Judging from recent industry trends, it would seem that companies have succeeded in implementing mass customization at the enterprise level. However, there are substantial difficulties in expanding mass customization strategies to the supply chain level. These difficulties are probably associated with tight requirements on supplier location and integration of information systems throughout the supply chain. Additionally, mass customization requires supply chain members to move to a higher level of trust and cooperation in order to address these problems jointly and in a mutually beneficial manner.

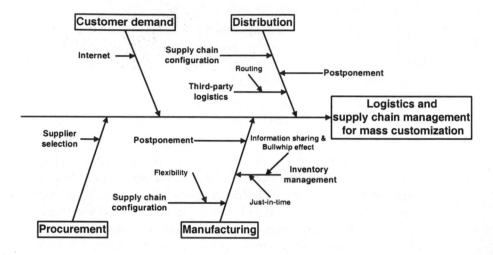

Figure 4-1. Key issues in logistics and supply chain management for mass customization

3. PROBLEM SOLVING APPROACHES: AN OVERVIEW

The main problem areas of logistics and supply chain management that are vitally important for mass customization have been identified in the previous section. This section reviews existing supply chain management models deemed suitable for addressing these problem areas with respect to contradicting mass customization objectives – low cost, large variety and shorter lead times. Utilization of these mechanisms allows for reducing logistics costs, lead times, and realizing improved responsiveness to customer demand. The closing sub-section describes binding among these problem areas.

3.1 Supply Chain Configuration

In the supply chain context, the facility location problem is to be considered across all tiers of the supply chain. These models are referred to as supply chain configuration or structure models. (Owen and Daskin (1998) review facility location problem in the framework of single-tier systems.)

The primary objective of supply chain configuration models is to determine location of suppliers, manufacturing facilities, distribution centers and to establish flows among supply chain members. The configuration decisions are generally long-term decisions. The main factors considered in these models include costs related to fixed investment in facilities, processing, procurement, transportation and capacity constraints. As can be seen, the supply chain configuration problem is also tightly coupled with the supplier selection problem. Therefore, these problems are discussed jointly in this Chapter.

There is not a good body of published research on the issue of supply chain configuration for mass customization. To adopt existing models for decision making in the mass customization framework, two major requirements – lead time and handling of possibly large product variety, are to be considered. The product variety can be handled by using data aggregations, which seems the natural choice because customized products can be associated with a product family. However, accuracy of data aggregations and its impact on network optimization remain one of the scarcely investigated problem areas (Ballou 2001).

Modeling steps involved in supply chain configuration are shown in Figure 4-2 (the figure is based on the supplier selection framework as proposed by De Boer et al. (2001)). Supplier selection criteria are elaborated at the first step. The two main criteria particularly important for mass customization are costs and time. It can be expected that the time criterion have a larger weight in the multi-criteria decision-making, than in the case of mass production. Quality of supplies also is essential as an enabler of the two main criteria. Qualification deals with initial reduction of the set of suppliers, identifying alternative locations for manufacturing facilities and distribution alternatives. The supply chain configuration is actually established at the choice step. Methods used in this step include linear weighting (Choy et al. (2002)) and mathematical programming (Tsiakis et al. (2001)). Although these methods allow for effective evaluation of a large number of alternative configurations, they have a limited scope and are difficult for representing dynamic and stochastic issues. The limited scope implies that relationships between configuration, and operational level decisions are not explored, even though these interactions can be substantial. These deficiencies are particularly important in the mass customization framework. Therefore, the detailed evaluation step has been added as the final stage of the supply chain configuration modeling. The detailed evaluation can be conducted by simulating performance of the established supply chain configuration at the operational level (the joint utilization of analytic and simulation models has been referred to as the hybrid approach (Shanthikumar and Sargent 1983), which allows obtaining comprehensive understanding of processes within the supply chain and consequences of adopting mass customization. One of the major problems associated with this approach is elaboration of a feedback mechanism between the choice and detailed evaluation steps.

Sabri and Beamon (2000) have presented one of the most complete developments in this area. They elaborate a multi-objective optimization model integrating a strategic supply chain reconfiguration model with analytic operational level models. Optimization objectives are cost minimization and volume flexibility maximization. The strategic level optimization model determines, which plants and distribution centers to open, assigns customers to the distribution centers and establishes product flows downstream starting with suppliers. The operational level models represent demand and operating uncertainties and provide feedback to the strategic model by updating its parameters. The operational level models are also used to estimate lead times, although the strategic model does not contain time constraints. Authors demonstrate that inclusion of the operational models has substantial impact on selection of supply chain configuration. However, the utilization of the analytic models at the operational level limits flexibility of the detailed evaluation.

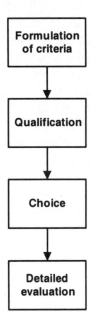

Figure 4-2. Supply chain configuration modeling steps

3.2 Postponement

Postponement is one of the dominant strategies used to address mass customization issues. The postponement strategy implies that differentiation of products regarding their form and place is delayed to the latest possible point in the supply chain (Bowersox and Morash 1989). Van Hoek (2001) classifies types of postponement and surveys related research. Postponement can take place in manufacturing by delaying final assembly, labeling and packaging. It also can take place in distribution by delaying decisions on committing products to a certain distribution area. Overlap between manufacturing and logistical postponement occurs by performing final manufacturing activities during the distribution phase, especially, in the context of global supply chains. Postponement is shown to have a limited value for providing full customization.

The number of delayed supply chain operations, which will be completed only after receiving a customer order, defines the level of postponement. Waller et al. (2000) analyze costs associated with using postponement and elaborate a model for selecting the optimal level of postponement. Postponement reduces the inventory carrying cost because less valuable components are stored instead of more valuable end products. The inventory carrying cost for the given level of postponement is higher, if the level of customization is higher. However, increasing the level of postponement can compensate for the increase of the inventory carrying cost due to the higher level of customization. The transportation cost is characterized as substantially dependent upon properties of supply chain and processed products. It can increase due to more frequent shipments, while a cost reduction can be achieved by improved efficiency of component transportation and using direct shipments to customers. Postponement incurs lost sales cost because some customers are not willing to wait. This cost also depends upon the level of postponement. Authors argue that lost sales cost can be reduced by providing higher level of customization, i.e., customers are likely to accept longer waiting times, if more custom features are offered.

A trade-off between the inventory cost and the delivery time according to the level of postponement for a particular manufacturing supply chain is analyzed in Figure 4-3 (Chandra and Grabis 2001). A multi-stage manufacturing supply chain has been simulated. Custom features are added to the distribution component of the supply chain. Manufacturing is initiated upon receiving customer orders at supply chain units downstream relative to the decoupling point. Observations are obtained by varying positioning of the decoupling point in the supply chain. The curve suggests that large reduction of the inventory cost accompanied by a relatively small increase of delivery time can be achieved by postponing the final assembly. Postponement of the sub-assembly results in insignificant reduction of the inventory cost and the substantial increase of the delivery time. The decreasing width of the 95% confidence intervals suggests that the postponement reduces inventory cost uncertainty.

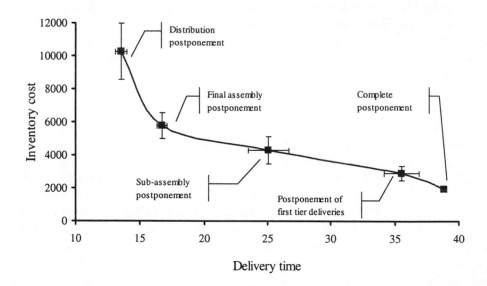

Figure 4-3. The trade-off between the inventory cost and the delivery time according to the level of postponement

Van Hoek (1998) and Van Hoek et al. (1999) have analyzed relationships between postponement and facility location by means of industry survey. Authors indicate that postponement allows reducing a number of distribution centers, while decisions on locating these distribution centers depend upon both transportation considerations and abilities of a particular area to support manufacturing and sourcing operations.

Postponement also has emerged as a value added service offered by third-party logistics providers (Van Hoek 2000). Offering packaging and final assembly services is expected to grow quickly. Therefore, third-party logistic providers will face the necessity to enhance their technological capabilities. Van Hoek (2000) proposes the framework for expanding third-party logistics services, supporting postponement. The framework describes the role of information technology and strategic partnerships in achieving higher supply chain integration, improved control and better customer service.

3.3   Inventory Management

The inventory management problem in the context of mass customization is a multi-stage, multi-item inventory management problem (see Graves et al. (1993) for presentation of multi-stage inventory management approaches and methods). Due to customization, demand for components can be unevenly distributed. Core components are demanded at a similar rate as in the case of mass production; most often used components may have steady demand, while less frequently demanded components might have erratic demand. Such components require different approaches to inventory management. MRP based policies can be used for managing components with variable demand. Re-order point policies can be used for managing globally sourced components with steady stochastic demand. JIT policies can be used for managing locally sourced components with steady demand. This division demonstrates interrelationships between inventory management and supply chain configuration. Another challenge is synchronization of inventory management for all components, in order to achieve a desirable service level for customized products. Most often, customized products are build-to-order and thus no end-product inventory is kept. The most relevant service measure is the delivery delay after offsetting the promised delivery time.

Ho and Chang (2001) describe a model combining JIT and MRP inventory management approaches. The combination allows using the MRP approach in the pull production mode. Even though JIT inventory policies have been advocated for utilization in the mass customization framework, there appear to be substantial limitations, such as requirements regarding supplier location and steady demand, and limited flexibility because JIT is perceived as a long-term commitment.

Soman et al. (2003) provide a condensed literature review of mixed make-to-order and make-to-stock systems. Authors identify parameters determining the choice between make-to-order and make-to-stock in the food industry.

Cheng et al. (2002) consider a configure-to-order system, where a product is configured from standardized components upon receiving a customer order. The system is argued to be an ideal approach to providing both mass customization and quick response. Contrary to the traditional assemble-to-order system, the configure-to-order system offers nearly unlimited number of product configuration. Inventory replenishment lead-time for components is relatively large. Authors develop a base-stock policy for managing components inventory with service level requirements, depending upon end-product configuration. The service level is measured by off-the-shelf availability of components.

Another alternative would be considering an inventory system with service level depending upon complexity of product configuration, where the service level is measured by delivery time. If a customer requires rarely demanded component, a longer delivery time is offered. This situation is similar to on-line bookstores, where some books are kept in stock, while others have longer delivery times. However, value of the demanded product, willingness of the customer to wait, and potential express deliveries of components should also be considered.

3.4   Third Party Logistics

Companies can outsource their logistic operations including material sourcing and product distribution to third-party logistics providers. Third-party providers can reduce costs of mass customization by 1) transporting and warehousing materials and products, and 2) providing add-on services, such as final customization.

Third-party logistics providers have elaborated transportation networks with high customer coverage. They offer their clients high level of flexibility in terms of delivery speed, packaging and technology. Adopting third-party logistics services allows companies to focus on their core competencies, provide access to established distribution networks and offer high level of flexibility. These are the main reasons companies seek to outsource logistics operations. Importance of these issues only increases in the mass customization framework. Disadvantages of third-party logistics are relative loss of control and potential conflict with core competencies, if logistics is one of them.

Bolumole (2001) identifies following drivers behind choosing third-party logistics services: 1) complex supply chain due to fragmented supply basis, 2) increasing volume of product returns, 3) need to improve service, 4) cost benefit through larger volumes, 5) access to existing infrastructure, 6) simplification of supply chain processes, 7) consistent delivery times, accuracy and efficiency, and 8) reduced overheads. The third-party logistics providers seek to improve utilization of their assets. The fourth and seventh drivers apparently are highly significant for mass customization. Distribution of customized products from the manufacturer perspective is often conducted on per item basis, making it difficult for the manufacturer to achieve an economic volume, while a third-party logistics provider achieves economic volume by combining order from several clients. Customers are likely to be more willing to wait for a product if delivery times are predictable and consistent.

Sink and Langley (1997) describe a five-stage methodology for selecting the third-party logistics provider. The first step is to identify need for logistics outsourcing; the second to identify feasible alternatives; the third to evaluate and select the provider; the fourth to implement the service; and the fifth and final step is ongoing service assessment. Menon et al. (1998) discuss criteria used by companies to select a third-party logistics provider. The most important criteria are on-time shipments and deliveries, meeting promises, availability of management and error rate. Authors suggest using separate evaluation stages for selecting the provider according to the performance criteria and price.

Tyan et al. (2003) develop a model for evaluation of strategies for freight consolidation by a global third-party logistics provider. This enables reducing shipment costs for individual contractors. However, it may cause increasing delivery lead times. Authors demonstrate how consolidation can be achieved without sacrificing speed of delivery. The developed model is a linear program formulation representing a particular consolidation policy. The best policy is determined by comparing three different linear programming models.

Third-party logistics providers also offer a range of supplementary services (Van Hoek 2000). The survey of third-party logistics providers shows that the most often offered supplementary services are inventory management and registration, consulting on logistics strategy, billing the customer and handling of returns. However, services generally are still offered at a rather low level.

Meade and Sarkis (2002) develop a conceptual model for selecting the third-party reverse logistics providers. Lewis and Talalayevsky (2000) discuss third-party logistics in the framework of inter-organizational information systems. Authors argue that third-party logistics providers act as supply chain integrators by providing standardized information exchange.

3.5 Information Technology

The role of information technology in mass customization, particularly in supply chain management for mass customization has been aptly summarized by Andel (2002), stating, "… mass customization requires mass communication". Primary contributions of information technology to supply chain management and logistics in the mass customization framework are receiving of customer orders, transformation of customer orders into manufacturing orders, and handling (including control) of information flows regarding inventory and transportation.

Yao and Carlson (1999) discuss the architecture of the real-time inventory management information system. All product handling operations are monitored using bar-code scanners from which information is transmitted to the warehouse control system using the radio-frequency communications in real-time. These real-time data can be used to improve inventory management decisions. The inventory management system is linked with suppliers and customers through the electronic data interchange (EDI). Advanced delivery orders sent using EDI allow both the warehouse operator and customers to improve planning of their activities, thus improving supply chain coordination. Main benefits of the real-time information system with regard to mass customization include reduced lead times, accurate and timely information about product and component availability, advance information about expected delivery time and higher shipment accuracy.

Lee et al. (1999) have analyzed impact of EDI on performance of supply chain participants showing that EDI benefits both its champions and adopters by increasing inventory turnaround and reducing stock outs.

The Internet has become the main tool for communicating customer requirements to manufacturer. Computer manufacturers allow users to configure a custom built computer using a simple Web based interface. For instance, on the Web site of one of the major computer manufacturers, after specifying the computer family (e.g. upscale laptop), the consumer selects modules from nine categories with multiple selections for each category. The number of possible configuration is of order 10^5. The order is to be assembled within 14 days, and promised shipment time is 2 to 7 days depending on the delivery mode. Besides directly linking the end-customer and manufacturer, the Internet also provides additional information, such as customer guides and product comparisons. More complex information systems for communicating between manufacturer and business customers may be needed.

The Delphi qualitative forecasting study conducted by Akkermans et al. (2003) shows that key issues of supply chain management in the coming years include integration of activities between suppliers and customers, flexibility of information technologies and increasing customization of products and services. It can be seen that these three issues are tightly interrelated because flexibility of information technologies facilitates cooperation among supply chain members, while the higher level of cooperation contributes to expanding mass customization capabilities. Akkermans et al. (2003) also analyze the expected impact of ERP systems on supply chains. Although the overall impact is forecasted to be only modest, the ERP systems are expected to substantially facilitate adoption of mass customization. These systems provide means for communicating with customers and eventual transformation of customized orders into appropriate production orders. However, the study indicates that communications could be improved by moving to an Internet-like architecture of ERP systems.

3.6   Links Among the Problems Areas

Supply chain configuration is one of primary decisions to be made before initializing supply chain operations. It has profound impact on many other forthcoming supply chain management decisions. Postponement and inventory management are tightly coupled. The location of the decoupling point depends upon the supply chain configuration established and the inventory management policy adopted. Different inventory policies can be adopted for using before and after product differentiation. The inventory management problem is solved independently of postponement in supply chains with limited utilization of postponement. Third-party logistics can offer a wide variety of services including inventory management. It also acts as one of the facilitators of postponement. Third-party logistics can be used as a tool for simplification of inventory management and many supply chain management activities. Finally, information technology supports execution of all logistics and supply chain management activities. With regard to the key problem areas covered, information technology enables efficient functioning of distributed configurations, efficient inventory management (especially, information sharing and coordination) and productive cooperation between supply chain stakeholders and third-party logistics providers.

4.   MODELS FOR IMPLEMENTING MASS CUSTOMIZATION POLICIES IN SUPPLY CHAIN

This section analyzes the supply chain configuration and inventory management problems in more details. The analysis is based on the conceptual supply chain configuration framework described in Chapter 3 of this book. This framework implies that strategic supply chain configuration decisions are subjected to evaluation at the operational level in order to assess impact of the configuration established on supply chain operations.

4.1   Strategic View

Two configuration models are presented. The first focuses on capacity planning under demand uncertainty, while the second focuses on impact of lead time on establishing the supply network. Conceptually, both models can be easily integrated to account for both uncertainty and lead time. However, computational feasibility is a concern. Therefore, both models are described separately and integration of results is left to the discretion of the decision maker.

4.1.1   Impact of Uncertainty

Capacity building decisions are needed in advance of actual production horizon, on the basis of uncertain demand information. Two categories of decisions can be identified: 1) how large facilities are needed, and 2) what products can be processed at which facilities. The second question is of particular importance in the case of mass customization because a high level of flexibility is needed. Flexibility is an approach for dealing with uncertainty. It enables varying quantitative production output and responding to changes in customer preferences regarding the type of product. A stochastic programming model solved using simulation-based optimization for capacity planning at flexible manufacturing facilities is developed (Chandra et al. 2003). Capacity requirements are determined by maximizing profit over the entire planning horizon subject to stochastic demand, product to plant assignments and several application specific constraints. These assignments are flexible within specific boundaries. The model is applied in the case study conducted at a major automotive OEM. The system considered consists of multiple plants and multiple products. Flexibility allows switching production from products with low demand to products with high demand.

The model-solving procedure is shown in Figure 4-4. The scenario loaded at the first step determines the level of flexibility in product to plant assignments provided. The capacity adjustment coefficients generated at the second step by a genetic optimization algorithm are used to alter the capacity requirements. A realization of stochastic demand is generated using the Monte-Carlo method at Step 3. At Step 4, a linear programming capacity allocation model is run given the current demand realization and the capacity level set at Step 2. The profit from selling the products is determined for the current demand realization (Step 5). The simulation procedure is replicated for multiple demand realizations. The expected profit for the given capacity level is found after simulation is completed at Step 6. If optimization is not completed, the process returns to Step 2 for the genetic algorithm to pick new values of the capacity adjustment coefficients. Otherwise, the optimization is completed and the optimal capacity requirements are selected.

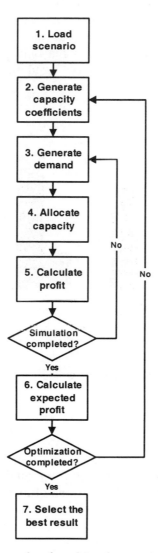

Figure 4-4. The procedure for solving the capacity planning model

Figure 4-5 shows capacity planning results for various levels of flexibility in product to plant assignments. Results are obtained using the optimization procedure for determining capacity requirements ("Optimized") and using judgmentally set capacity adjustment coefficients ("Judgmental"). Flexibility levels are referred as marginal, standard and higher. The profit is given relative to profit achieved for the case with the marginal flexibility level and the judgmentally set capacity adjustment coefficients. Profitability improves with an increasing level of flexibility in both cases. However, optimization of capacity requirements has led to substantially improved profitability. These results indicate that increased flexibility needed for producing large product variety is not the only precondition for successful operations; optimization of capacity requirements subject to demand uncertainty is also essential.

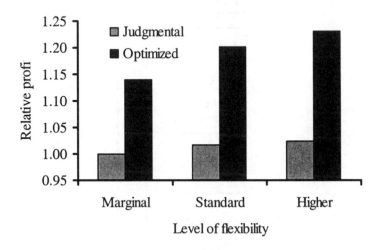

Figure 4-5. Relative profit according to the level of flexibility

4.1.2   Impact of Lead Time

Lead time reduction is one of the key issues in mass customization as discussed previously. There are multiple opportunities for lead time reduction. Reduction of procurement lead time is one of these opportunities. However, a manufacturer should not pursue the supply lead time reduction at any cost because other objectives such as low cost and high quality could be violated. Unfortunately, multi-criteria optimization of cost and lead time is difficult and often involves judgmental decisions. Here, an approach integrating optimization and simulation is proposed for optimizations of the supply network with regard to both cost, and supply lead time (Chandra and Grabis 2003). Simulation is used to evaluate costs associated with supply lead time. Afterwards, these costs are incorporated in the supplier selection optimization model, thus achieving optimization with respect to both procurement cost and supply lead time.

The problem is analyzed from the perspective of a manufacturer. The manufacturer has a choice among several alternative suppliers. The choice is made to minimize purchasing costs at the strategic level. However, the supply lead time provided by suppliers is also perceived as an important parameter because it influences accuracy of demand forecasts used in production planning. It is assumed to be the only supplier selection dependent parameter influencing the manufacturer at the operational level. Therefore, simulation results, which can be used to evaluate the true impact of lead time on manufacturer's total costs, should be obtained only for scenarios with different values of supply lead time regardless of actual suppliers selected. These results can be obtained prior to the supplier selection by optimization.

The objective function of the supplier selection model can be expressed as

$$TC = C_S + C_L \rightarrow \min ,$$

where TC is the total cost, C_S is the strategic level costs (e.g. material cost, transportation cost) and C_L is the lead time cost. The total cost is optimized subject to material requirements, supplier capacity and other application specific constraints. Two constraints for incorporating the simulation results are

$$l_i W_i + l < L, \ i = 1,..., M'$$

$$C_L = a + bL, \quad (1)$$

where l_i is the supply lead time provided by suppliers, l is the manufacturing lead time, L is the decision variable representing the cumulative lead time, which depends upon suppliers selected and M' is the number of suppliers. Constraint (1) is a meta-model for evaluation of the lead time cost according to the cumulative lead time. In order to construct the meta-model, the supply lead time cost is simulated for a few appropriately selected values of the supply lead time. The regression based meta-model is fitted using these values (see Kleijnen (1987) for developing meta-models according to simulation results). It is used to compute the lead time cost for all other values of the cumulative lead time; a and b are coefficients of the meta-model.

The simulation model represents inventory management for materials and sub-assemblies the manufacturer produces as base for further customizable products. It is assumed that at the operational level the manufacturer follows the build-to-stock policy and manages inventory using the MRP based inventory policy. Inventory management decisions are made according to short-term demand forecasts. The forecasting horizon depends upon both the supply lead time and the manufacturing lead time because the MRP based inventory policy normally requires freezing of the inventory replenishment schedule during the cumulative lead time (Yeung et al. 1998). Therefore, longer supply lead times increase forecasting horizon, which is associated with declining forecasting accuracy and reducing cost efficiency of inventory management.

The lead time cost is expressed as the manufacturer's inventory management cost

$$C_L = \sum_{t=1}^{T} (uI_{kt} + vY_{kt} + zB_{kt}), \quad (2)$$

where, T is the production horizon, u_t is the inventory holding cost, v_t is the fixed ordering cost and z_t is the lost sales penalty, I_{kt} is the end of period inventory, Y_t is the binary variable indicating whether an inventory replenishment order has been placed at period t and B_t is the quantity of units short. This cost depends upon the kth value of the cumulative lead time, but not on a particular supplier selection. The cost expression does not include inventory management cost for suppliers. Expression (2) is evaluated using simulation for selected values of the cumulative lead time L. Simulation is performed by generating short-term demand patterns and their forecasts and elaborating appropriate inventory replenishment schedules in the rolling horizon environment (for more details on modeling and simulating MRP based inventory systems see Yeung et al. (1998)).

The supplier selection model with accounting for the lead time cost (modified model) and the supplier selection model without accounting for the lead time cost (standard model) are compared in order to evaluate the importance of including operational factors in strategic decision making. The supply lead time for four alternative suppliers ($l_i, i = 1,...,4$), the manufacturing lead time (l), properties of the short-term demand process, namely, the autocorrelation coefficient (ρ) and the average demand to noise ratio (μ/σ), and the fixed ordering cost (v) are treated as variable experimental factors. The cost difference between the standard model and the modified model ΔTC is a dependent variable. A regression model is developed for experimental results obtained, in order to identify factors favoring use of the modified model. The fitted regression model with p-values for coefficients shown in parenthesis is (input data are standardized to range between 0 and 1 and cost difference is expressed in 10000's)

$$\Delta TC = \underset{(0.00)}{5.03} + \underset{(0.00)}{13.41}l_1 - \underset{(0.00)}{5.08}l_2 - \underset{(0.00)}{2.41}l_3 - \underset{(0.00)}{1.23}l_4$$

$$+ \underset{(0.99)}{0.01}l + \underset{(0.00)}{1.75}v + \underset{(0.00)}{14.37}\rho - \underset{(0.00)}{14.62}\overline{D}\Big/\sigma$$

Two of the most significant factors influencing differences between models characterize properties of the demand process. Stronger autocorrelation elevates the importance of forecasting and inventory planning, thus causing larger differences between the standard model and the modified model. Impact of the signal to noise ratio is similar. Lower values of the ratio cause larger forecasting errors and subsequently higher safety stock requirements, especially if lead times are long. The setup cost is the significant factor but the confidence level is lower than other significant factors. The manufacturer lead time is not a significant factor because it affects both models equally. Dependence of the cost difference according to the parameters of the short term demand process are shown in Figure 4-6. Results suggest that accounting for the operational factors is important, if the demand process is highly autocorrelated and noisy.

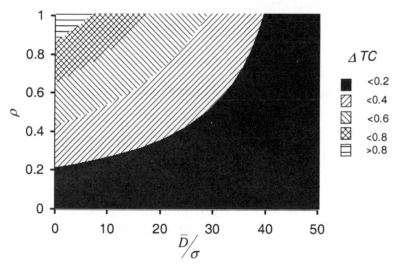

Figure 4-6. The contour graph of cost differences

4.2 Operational View

The operational level models used for evaluation of supply chain configuration decisions can be modified in numerous ways. This sub-section considers operational models for evaluation of the impact of forecasting on supply chain inventory performance and coordination among supply chain units. Even though mass customizers often have a luxury of knowing the end-customer demand, there is a need for forecasting in order to have materials and sub-assemblies ready to enable quick response to customer requests.

A simple supply chain inventory management problem is used to demonstrate importance of cooperation in inventory management. It is known that an information distortion phenomenon referred as the bullwhip effect is commonly observed in multi-tiered supply chains (Lee et al, 1997). This distortion obstructs efficient inventory management. A joint application of a proposed MRP based inventory management policy and autoregressive forecasting models is advocated for supply chains facing serially correlated demand to reduce negative impact of the bullwhip effect and at the same time to maintain high inventory performance for the most downstream unit (Chandra and Grabis 2002).

Traditional order-up policies place orders to replenish the stock depleted during the delivery lead time and the order size q_t is expressed as

$$q_t = \hat{D}_t^L - \hat{D}_{t-1}^L + D_{t-1},$$

where $\hat{D}_t^L$ is the estimated demand during L period long replenishment lead time, and D_{t-1} is the actual demand at the tth time period, respectively. The estimated demand is obtained using autoregressive forecasting models. This approach implies that an order placed at time t arrives at time $t+L$. This quantity will be used to satisfy the external demand for time periods starting from $t+L$. However, the ordered quantity is estimated using forecasts for time periods between t, inclusive, and $t+L$ exclusive. As a result, there is a disagreement between the period, for which consumption is forecasted, and the period in which consumption will occur. Such approach mainly relies on risk pooling rather than accuracy of forecasting. If there are predictable patterns in demand for sub-assemblies, then it is possible to order as many items as are expected to be consumed after arrival of the order. This ordering pattern is referred to as the MRP based approach. The order size is determined on the basis of forecast for the period $t+L$, achieved by substituting the lead time demand forecasts with appropriate ($L+1$)-steps forecasts obtained using autoregressive forecasting models.

$$q_t = \hat{D}_{t-1}(L+1) - \hat{D}_{t-L-2}(L+1) + D_{t-1}$$

$$= \mu + \sum_{i=1}^{\infty} \psi_{L+i} a_{t-i} - \sum_{i=1}^{\infty} \psi_{L+i} a_{t-L-i-1} + \sum_{i=0}^{\infty} \psi_i a_{t-i-1} , \quad (3)$$

$$= \mu + \sum_{i=0}^{L} (\psi_{L+i+1} + \psi_i) a_{t-i-1} + \sum_{i=L+1}^{\infty} \psi_{i+1} a_{t-i-1}$$

where μ is a non-negative constant, a_t represents the i.i.d. noise process with zero mean and finite variance σ^2 and ψ_i, $i=0,\ldots, \infty$, are random shock coefficient characterizing the serially correlated demand process (see Box et al. (1994) on modeling of autoregressive processes). Note that the actual demand during the previous period is compared with its multiple-steps forecast made at time $t-L-2$.

The order size expression (3) implies that the order size is equal to forecasted demand at the time of order arrival plus an adjustment for a forecasting error at the previous period. The forecasting horizon is $L+1$ because order is placed at time t for time $t+L$ and the last observed value of the external demand is D_{t-1}.

The order-up approach and the MRP-based approach are compared with respect to the bullwhip effect, and the inventory performance criterion, respectively. For purposes of comparison, the order variance, which compared with the demand variance characterizes the bullwhip effect, for both approaches is found. The order variance for the order-up approach is

$$var(q_t) = var\left[\mu + (C_1 + \psi_0) a_{t-1} + \sum_{i=1}^{\infty} (C_{i+1} - C_i + \psi_i) a_{t-i-1} \right],$$

$$= \sigma^2 (C_1 + \psi_0)^2 + \sigma^2 \sum_{i=1}^{\infty} \psi_{L+i}^2$$

where $C_i = \sum_{j=1}^{L} \psi_{i+j-1}$.

The order variance for the MRP-based approach is

$$var(q_t) = var\left[\mu + \sum_{i=0}^{L} (\psi_{L+i+1} + \psi_i) a_{t-i-1} + \sum_{i=L+1}^{\infty} \psi_{i+1} a_{t-i-1} \right]$$

$$= \sigma^2 \sum_{i=0}^{L} (\psi_{L+i+1} + \psi_i)^2 + \sigma^2 \sum_{i=L+1}^{\infty} \psi_{i+1}^2$$

In both cases, the order variance is larger than the demand variance for the positively correlated first order autoregressive processes, indicating presence of the bullwhip effect for both methods. However, numerical evaluation of the order variance relative to the demand variance shows that the MRP-based approach generally allows reducing the bullwhip effect compared to the order-up approach (Figure 4-7). At the same time, both methods give practically identical inventory performance measured as the ratio between the items in stock and items demanded (not shown). These results indicate that the MRP-based approach combined with autoregressive forecasting models allows improving coordination in supply chains by reducing the bullwhip effect without sacrificing the inventory performance.

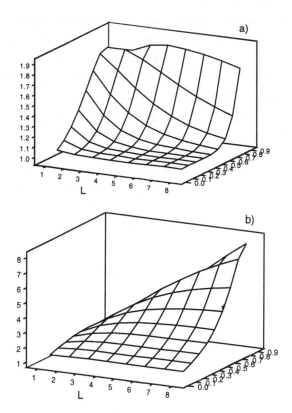

Figure 4-7. The bullwhip effect as the function of lead time and autocorrelation strength of the serially correlated first order demand process for a) the MRP based approach and b) the order-up approach

5. FUTURE RESEARCH DIRECTIONS

Table 4-1 summarizes problematic issues of logistics and supply chain management in the mass customization framework. It can be observed that many of these issues overlap several problem areas. Therefore, models providing an integrated analysis of supply chain management problems for implementing mass customization policies are required.

Table 4-1. Problematic issues of logistics and supply chain management for mass customization

Problem area	Issues
Supply chain configuration	Direct representation of dynamic and stochastic factors in supply chain configuration models or improved integration of these models with simulation models
Third-party logistics	Evaluation of supply chain configurations in the presence of third-party logistics providers Extent to which logistics operations should be outsourced
Postponement	Incorporating the level of postponement in supply chain configuration decisions
Inventory	Managing inventory of components, which have largely different demand patterns Evaluation of inventory policies during the supply chain configuration modeling stage
Information technology	Specialized policies for mass customization situations Internet based customer order placement and product design systems Integration of information systems between suppliers, manufacturers and business customers

In many cases, supply chains using mass customization policies tend to have a simple structure in order to speedup the operations. Insufficient collaboration between supply chain units is another reason preventing establishment of more complex supply chain configurations. Third-party logistics allows simplifying the supply chain configuration. However, competitive forces are likely to force increased supply chain configurations complexity. Therefore, the scope and comprehensiveness of supply chain configuration models is to be improved.

To the knowledge of authors, there are no published works on supply chain configuration, if distribution functions are outsourced to third-party logistics providers or a choice between a firm's own distribution network and third-party logistics providers is to be made. Despite the convenience of utilizing third-party logistics services, users of these services risk loosing a valuable business opportunity to other companies. Little research has been done on the evaluation of strategic value of logistic operations and a choice between third-party logistics providers and proprietary distribution networks.

Similarly, postponement decisions are tightly interrelated with supply chain configuration decisions. The current research mainly has focused on analyzing impact of postponement on inventory costs and customer service. However, if some of the manufacturing operations, other than packaging and labeling are postponed till distribution, substantial investments in the supply chain structure may be required. A model for simultaneous evaluation of postponement strategies and supply chain configurations is needed.

Supply chain configuration and different demand patterns for components cause differing requirements for managing inventory of components, used for producing customized products. Models for comparison and simultaneous utilization of different inventory management approaches are needed. Additionally, supply chain configuration decisions depend upon the inventory management approach selected. For instance, a configuration comprising a manufacturer receiving JIT supplies from local suppliers is to be compared with a configuration comprising a manufacturer receiving supplies according to a re-order point policy from distant suppliers. Currently, inventory management for mass customization relies upon methods developed for general inventory management situations. Mass customization brings additional inventory management parameters into the model. Inventory policies relating the level of customization and the customer service level is an interesting research area to be pursued.

Information technology supporting supply chain management for mass customization has reached a relatively high level of maturity, while mass customization still requires a higher level of integration of information systems among supply chain members.

REFERENCES

1. Ahlstrom P., Westbrook R. Implications of mass customization for operations management - An exploratory survey. International Journal of Operations & Production Management 1999; 19: 262-274.
2. Akkermans H. A., Bogerd P., Yucesan E., van Wassenhove L. N. The impact of ERP on supply chain management: Exploratory findings from a European Delphi study. European Journal of Operational Research 2003; 146: 284-301.
3. Alford D., Sackett P., Nelder G. Mass customisation -- an automotive perspective. International Journal of Production Economics 2000; 65: 99-110.
4. Andel T. From common to custom: The case for make-to-order. Material Handling Management 2002; 24-31.
5. Baarends E., Witbreuk J. M. A. Logistieke dienstverleners en verlanders bundelen hun krachten. Tijdschrift voor Inkoop en Logistiek 1994; 10: 32-36 (in Dutch).
6. Ballou R. H. Unresolved issues in supply chain network design. Information Systems Frontiers 2001; 3: 417-426.

7. Bolumole Y. A. The supply chain role of third-party logistics providers. The International Journal of Logistics Management 2001; 12: 87-102.
8. Bowesox D. J., Morash E. A. The integration of marketing flows in channels of distribution. European Journal of Marketing 1989; 23: 58-67.
9. Box G. E. P., Jenkins G. M., Reinsel G. C., Time Series Analysis: Forecasting and Control. Englewood Clifs: Prentice-Hall, 1994.
10. Chandra, C., Grabis, J. Reconfiguration of multi-stage production systems to support product customization using generic simulation models. Proceedings of the 6th International Conference of Industrial Engineering Theory, Applications and Practice, November 18-20, San-Francisco, CA, 2001.
11. Chandra, C., Grabis, J. Application of multiple-steps forecasts to restrain the bullwhip effect, *Working paper*, University of Michigan-Dearborn, 2002.
12. Chandra C., Grabis J. Impact of supply lead time on supplier selection: An operational perspective. Proceedings: Twelfth Annual Industrial Engineering Research Conference (IERC), May 18-20, Portland, Oregon, 2003.
13. Chandra, C., Everson, M., Grabis, J. Evaluation of enterprise-level benefits of flexibility, Working paper, University of Michigan-Dearborn, 2003.
14. Cheng F., Ettl M., Lin G., Yao D. D. Inventory-service optimization in configure-to-order systems. Manufacturing & Service Operations Management 2002; 4: 114-132.
15. Choy K. L., Lee W. B., Lo V. An intelligent supplier management tool for benchmarking suppliers in outsource manufacturing. Expert Systems with Applications 2002; 22: 213-224.
16. Da Silveira G., Borenstein D., Fogliatto F. Mass customization: Literature review and research directions. International Journal of Production Economics 2001; 72: 1-13.
17. De Boer L., Labro E., Morlacchi P. A review of methods supporting supplier selection. European Journal of Purchasing & Supply Management 2001; 7: 75-89.
18. Dowlatshahi S. A modeling approach to logistics in concurrent engineering. European Journal of Operational Research 1999; 115: 59-76.
19. Eastwood M. A. Implementing mass customization. Computers in Industry 1996; 30: 171-174.
20. Erens F. J., Hegge H. M. H. Manufacturing and sales co-ordination for product variety. International Journal of Production Economics 1994; 37: 83-99.
21. Feitzinger E., Lee H. L. Mass customization at Hewlett-Packard: The power of postponement. Harvard Business Review 1997; 75: 116-121.
22. Fisher M. L., Ittner C. D. The impact of product variety on automobile assembly operations: Empirical evidence and simulation analysis. Management Science 1999; 45: 771.
23. Furst K., Schmidt T. Turbulent markets need flexible supply chain communication. Production Planning & Control 2001; 12: 525-533.
24. Ghiassi M., Spera C. Defining the Internet-based supply chain system for mass customized markets. Computers & Industrial Engineering 2003; 45: 17-41.
25. Gooley T. B. Mass customization: How Logistics makes it happen. Logistics 1998; 4: 49-53.
26. Graves S. C., Rinnooy Kan A. H. G., Zippkin P. H., Handbooks in OR & MS, Logistics of Production and Inventory. Amsterdam: North Holland, 1993.
27. Ho J. C., Chang Y.-L. An integrated MRP and JIT framework. Computers & Industrial Engineering 2001; 41: 173-185.
28. Jiao J., Ma Q., Tseng M. M. Towards high value-added products and services: mass customization and beyond. Technovation 2003; 23: 809-821.
29. Kleijnen J. P. C., Statistical tools for simulation practitioners. New York: Marcel Dekker, 1987.

30. Kotha S. From Mass Production to Mass Customization: The Case of the National Industrial Bicycle Company of Japan. European Management Journal 1996; 14: 442-450.
31. Lambert D. M., Cooper M. C. Issues in Supply Chain Management. Industrial Marketing Management 2000; 29: 65-83.
32. Lambert D. M., Stock J. R., Ellram L. M., Stockdale J., Fundamentals of Logistics. New York: McGraw-Hill/Irwin, 1997.
33. Lee H. G., Clark T., Tam K. Y. Research report. Can EDI benefit adopters? Information Systems Research 1999; 10: 186-195.
34. Lee H. L., Padmanabhan V., Whang S. Information distortion in a supply chain: The bullwhip effect. Management Science 1997; 43: 546.
35. Lewis I., Talalayevsky A. Third-party logistics: Leveraging information technology. Journal of Business Logistics 2000; 21: 173-185.
36. Meade L., Sarkis J. A conceptual modelfor selecting and evaluating third-party reverse logistics providers. Supply Chain Management: An International Journal 2002; 7: 283-295.
37. Menon M. K., McGinnis M. A., Ackerman K. B. Selection criteria for providers of third-party logistics services: An exploratory study. Journal of Business Logistics 1998; 19: 121-137.
38. Mentzer J. T., DeWitt W., Keebler J. S., Min S., Nix N. W., Smith C. D., Zacharia Z. G. Defining supply chain management. Journal of Business Logistics 2001; 22: 1-25.
39. Nevins J. L., Whitney D. E., Concurrent Design of Products and Processes. In ed. New York: McGraw-Hill, 1989; 2-3.
40. Owen S. H., Daskin M. S. Strategic facility location: A review. European Journal of Operational Research 1998; 111: 423-447.
41. Rabinovich E., Dresner M. E., Evers P. T. Assessing the effects of operational processes and information systems on inventory performance. Journal of Operations Management 2003; 21: 63-80.
42. Randall T., Ulrich K. Product Variety, Supply Chain Structure, and Firm Performance: Analysis of the U.S. Bicycle Industry. Management Science 2001; 47: 1588-1604.
43. Sabri E. H., Beamon B. M. A multi-objective approach to simultaneous strategic and operational planning in supply chain design. Omega 2000; 28: 581-598.
44. Shanthikumar J. G., Sargent R. G. A Unifying View of Hybrid Simulation/Analytic Models and Modeling. Operations Research 1983; 31: 1030-1052.
45. Silver E. A., Pyke D. F., Peterson R., Inventory management and production planning and scheduling. New York: John Wiley & Sons, 1998.
46. Sink H. L., Langley C. J. A managerial framework for the acquisition of third-party logistics services. Journal of Business Logistics 1997; 18: 163-189.
47. Slats P. A., Bhola B., Evers J. J. M., Dijkhuizen G. Logistics chain modelling. European Journal of Operational Research 1995; 87: 1-20.
48. Soman C. A., Van Donk D. P., Gaalman G. Combined make-to-order and make-to-stock in a food production system. International Journal of Production Economics 2003 (In Press).
49. Tsiakis P., Shah N., Pantelides C. C. Design of multi-echelon supply chain networks under demand uncertainty. Industrial & Engineering Chemistry Research 2001; 40: 3585-3604.
50. Tyan J. C., Wang F.-K., Du T. C. An evaluation of freight consolidation policies in global third party logistics. Omega 2003; 31: 55-62.
51. Van Hoek R. I. Reconfiguring the supply chain to implement postponed manufacturing. International Journal of Logistics Management 1998; 9: 95-110.
52. Van Hoek R. I. The role of third-party logistics providers in mass customization. The International Journal of Logistics Management 2000; 11: 37-46.

53. Van Hoek R. I. The rediscovery of postponement a literature review and directions for research. Journal of Operations Management 2001; 19: 161-184.

54. Van Hoek R. I., Vos B., Commandeur H. R. Restructuring European supply chains by implementing postponement strategies. Long Range Planning 1999; 32: 505-518.

55. Verwoerd W. Value-added logistics: The answer to mass customization. Hospital Materiel Management Quarterly 1999; 21: 31-36.

56. Waller M. A., Dabholkar P. A., Gentry J. J. Postponement, product customization, and market oriented supply chain management. Journal of Business Logistics 2000; 21: 133-159.

57. Yao A. C., Carlson, John G. The impact of real-time data communication on inventory management. International Journal of Production Economics 1999; 59: 213-219.

58. Yeung J. H. Y., Wong W. C. K., Ma L. Parameters affecting the effectiveness of MRP systems: a review. International Journal of Production Research 1998; 36: 313-331.

SECTION 3:

SUPPORTIVE TECHNIQUES AND TECHNOLOGIES FOR ENABLING MASS CUSTOMIZATION

CHAPTER 5

A DECOMPOSITION METHODOLOGY FOR UNCOUPLED MODULAR PRODUCT DESIGN

Ali K. Kamrani[1], Sa'ed M. Salhieh[2]

[1]*University of Houston*
[2]*The University of Jordan*

Abstract: Before the application of the modularity concept, the design process involved developing a unique and optimum design for a given product or system. Designers focused on satisfying a set of design and manufacturing attributes while maintaining a predetermined target cost. As a result, a complex design was proposed, incorporating functionally interdependent layers, requiring costly and time consuming iterations. This process provided highly coupled designs, where modification of one area directly affects other areas of design. This chapter addresses modularity and its use in un-coupled approach for product design.

Keywords: Modularity, un-coupled design, design axioms, product development.

1. INTRODUCTION

Product development contains a set of ongoing activities that an organization must perform to develop, manufacture, and a sell a product (Pugh 1991, Ulrich and Eppinger 2000). These activities are the result of a multidisciplinary effort that includes marketing, research, engineering design, quality assurance, manufacturing, and a whole chain of suppliers and vendors. Furthermore, product development comprises all strategic planning, capital investments, management decisions and tasks necessary to create a new product. At the heart of the development process is the *engineering design process,* that is the process of devising a system, component, or process to meet desired needs (Pahl and Beitz 1996). Most product development methodologies share a common goal; to satisfying customer needs by incorporating the voice of the customer. Furthermore, the competitive nature of today's market pressured organizations to reduce development time and cost while maintaining high quality. This lead to the need to develop new methodologies that has the ability to identify customer requirements and use those requirements to rapidly build products.

This chapter presents a new systematic approach to design decomposition. The proposed steps allows for a designer to transform a coupled design into a set of uncoupled independent designs that could be developed in parallel. It also illustrates how "Quality Function Deployment" and "Design for Modularity" are combined to construct a development map that begins by understanding the voice of the customer, and proceed to product feature resulting in products that could be developed independently of each other.

2. CUSTOMER SATISFACTION

Successful product development can be achieved by targeting the development effort towards producing products that can satisfy customers by meeting their expectations. The focus on customer satisfaction is a direct result of the competitive nature of current global marketplace. Satisfying customers can be viewed as an ongoing process where organizations need to maintain continuous communication channels with the customer in order to understand their true wants and needs (Finkelman and Goland 1990); furthermore organizations need to collect customers' feedback and reactions toward new products and incorporate these reactions into the product development process to improve products (Kamrani and Salhieh 2002). Kano et al. (1984) presented a conceptualization model of customer satisfaction (Figure 5-1) (Besterfield et al 1995).

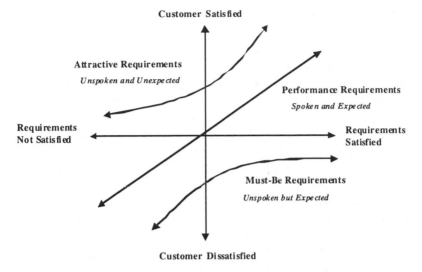

Figure 5-1. Kano's Model

In this model, three major types of product requirements that affect customer satisfaction are recognized. The "must-be" requirements constitute the main reason that the customer needs the product. The customer expects these requirements, and the manufacturer gets no credit if they are there. The "performance" requirements are spoken needs. They can be measured for importance as well as for range of fulfillment levels. Customers expect that the product will meet these requirements. Finally the "attractive" requirements are future oriented and usually high-tech innovations. These requirements are unspoken and unexpected because the customer does not know they exist. Usually, these requirements come as a result of the creativity of the research and development effort in the organization. Another major concept that support the incorporation of the voice of customer into the planning and design process is the Quality Function Deployment. QFD is a team-based technique that provides a mechanism for identifying and translating customer requirements into technical specifications for product planning, design, process, and production (Matzler and Hinterhuber 1998, Shillito 1994). In general, quality function deployment addresses a) customer wants and needs, b) importance of the requirements to the customer and 3) product features needed to satisfy the customer's needs. Implementing QFD relies on using the house of quality as a conceptual map that provides the means for cross-functional planning and communication (Sullivan 1986).

3. MODULAR DESIGN

The concept of modularity provides the necessary foundation for organizations to design products that can respond rapidly to market needs and allow the changes in product design to happen in a cost-effective manner. Modular products are products that fulfill various overall functions through the combination of distinct building blocks or modules, in the sense that the overall function performed by the product can be divided into sub-functions that can be implemented by different modules or components (Salhieh and Kamrani 1999, Shirly 1992). An important aspect of modular products is the creation of a basic core unit to which different components (modules) can be fitted, thus enabling a variety of versions of the same module to be produced. The core should have sufficient capacity to cope with all expected variations in performance and usage. Components used in a modular product must have features that enable them to be coupled together to form a complex product. Designing a modular product can be done by using conventional product development techniques, but using these techniques will not lead to a reduction in product development lead time, and thus a new development methodology is needed that can utilize the full strength of the modular architecture of products. Using the concept of modularity in product design focuses on decomposing the overall design problem into functionally independent sub-problems, in which interaction or interdependence between sub-problems is minimized. Thus, a change in the solution of one problem may lead to a minor modification in other problems, or it may have no effect on other sub-problems. That is, the modular design concept attempts to establish a design decomposition technique that reduces the interaction between design components (or modules) to reduce the complexity and development time of a product. A modular design usually is adaptable with little or no modification for many applications (Kamrani and Salhieh 2002, Ulrich and Tung 1991). Modular design can also be viewed as the process of first producing units that perform discrete functions, then connecting the units together to provide a variety of functions. Modular design emphasizes the minimization of interactions between components, enabling components to be designed and produced independently from each other. Each component designed for modularity is supposed to support one or more functions. When components are structured together to form a product, they will support a larger or general function. This shows the importance of analyzing the product function and decomposing it into sub-functions that can be satisfied by different functional modules.

4. AXIOMATIC DESIGN

A concept widely used for the development of modularity is the axiomatic design (Suh 1995). Axiomatic design theory is a visual way of expressing the design intent and objectives. Two important elements that must be defined before using the axiomatic design are:

1. **Functional Requirement:** Minimum set of unique requirements that describes the design objectives for specific needs. FRs are independent of each other at every level of hierarchy.
2. **Design Parameter:** The set of physical entities that are designed to address FRs.

Functional requirements (Whats) are the response to the list of the customer requirements (CR), and design parameters (Hows) are the requirements for proper design process. Figure 5-2 illustrates the four domain associated with the design process.

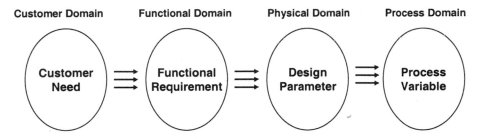

Figure 5-2. Design domains

At any given hierarchy, the functional requirements that define the design intent are illustrated by vector {FRs}. The physical characteristics that are the responses to the FRs are represented using vector {DPs}. The mathematical representation of the axiomatic design can now be defined using matrix notation shown below (Suh 1995):

$$\{FRs\} = [A] \{DPs\}$$

[A] is defined as the design matrix. The two major design axioms are:
1. **Independence Axiom**: An optimal design always maintain the independence of the functional requirements.

2. **Information Axiom**: The best design is a functionally uncoupled design that has minimum information content.

The mathematical representation of a coupled design (concurrent design) and the scope are illustrated in Figure 5-3.

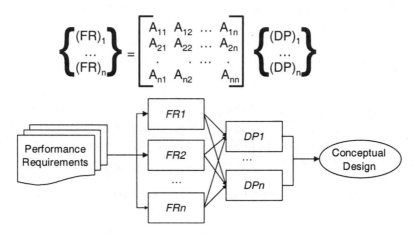

$$\left\{ \begin{array}{c} (FR)_1 \\ \cdots \\ (FR)_n \end{array} \right\} = \begin{bmatrix} A_{11} & A_{12} & \cdots & A_{1n} \\ A_{21} & A_{22} & \cdots & A_{2n} \\ \cdot & & \cdot & \cdot \\ A_{n1} & A_{n2} & & A_{nn} \end{bmatrix} \left\{ \begin{array}{c} (DP)_1 \\ \cdots \\ (DP)_n \end{array} \right\}$$

Figure 5-3. Coupled design interrelationship

In coupled design, each FRs directly relates to every DPs. If matrix [A] is transformed into either a triangular or diagonal form, the independence axiom is satisfied. These are known as decoupled (quasi-coupled) and uncoupled designs. In decoupled design or sequential design, a FR is related to one or more DPs. The mathematical representation and the scope of the decoupled design are shown in Figure 5-4.

An uncoupled or parallel design is the case where specific FRs are addressed by only one DP. The mathematical representation and the scope of uncoupled design are shown in Figure 5-5.

Using matrix decomposition theory, optimum design parameters could be identified that directly satisfy the defined FRs.

$$\begin{Bmatrix} (FR)_1 \\ \cdots \\ (FR)_n \end{Bmatrix} = \begin{bmatrix} A_{11} & 0 & \cdots & 0 \\ A_{21} & A_{22} & \cdots & 0 \\ \cdot & \cdot & \cdots & \cdot \\ A_{n1} & A_{n2} & & A_{nn} \end{bmatrix} \begin{Bmatrix} (DP)_1 \\ \cdots \\ (DP)_n \end{Bmatrix}$$

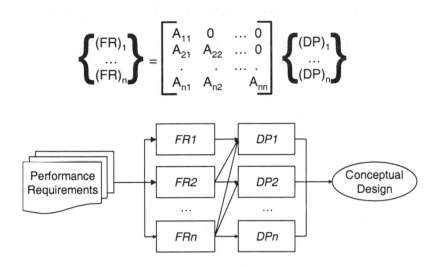

Figure 5-4. Decoupled design interrelationship

$$\begin{Bmatrix} (FR)_1 \\ \cdots \\ (FR)_n \end{Bmatrix} = \begin{bmatrix} A_{11} & 0 & \cdots & 0 \\ 0 & A_{22} & \cdots & 0 \\ \cdot & \cdot & \cdots & \cdot \\ 0 & 0 & & A_{nn} \end{bmatrix} \begin{Bmatrix} (DP)_1 \\ \cdots \\ (DP)_n \end{Bmatrix}$$

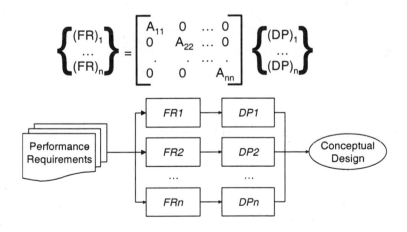

Figure 5-5. Uncoupled design interrelationship

5. METHODOLOGY FOR DECOMPOSITION AND ANALYSIS FOR MODULAR PRODUCT DESIGN

The overview of the proposed model is illustrated in Figure 5-6. The methodology is divided into four major phases.

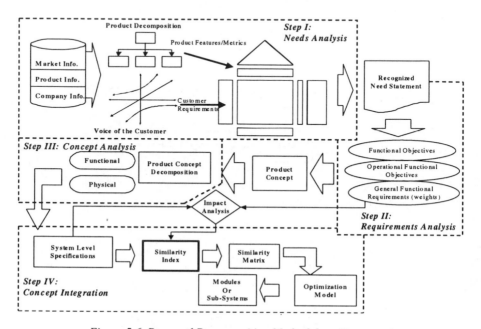

Figure 5-6. Proposed Decomposition Methodology Framework

5.1   Needs Analysis and Prioritization

Several sources of information could be used to identify the needs. Such sources include potential customers, the firm for which the design is being made, the competition, and any authorities that can impose restrictions on the product (standards, safety, etc.). Information sources can be grouped into two large groups: existing information and new information as in Table 5-1. The information collected are analyzed and then organized according to Kano's model.

Table 5-1. Information Sources

Existing Information	New Information
• Sales Data • Customer Complaints • Warranty Data • Publications from the government and trade journals • The firm's designers, engineers and managers • Benchmarked Products	• Surveys, including mail, telephone, comment cards • Focus Groups • Interviews, group or individual • Service calls and trade shows • Direct visits with the users

These collected information are further categorized as:

a. Expected Requirements

These are the basic requirements that customers expect to exist self-evidently. These requirements are satisfied through basic product/service characteristics that customers assume are part of the product or service; that is, they expect them as standard feature.

b. Unspoken Requirements

Product features that customers do not talk about and, though silent, are important and cannot be ignored. Unspoken requirements are like expected requirements in the sense that customers do not feel obligated to tell the developer about them, either because they feel they are clear and obvious or because they do not know that they exist.

c. Spoken Requirements

Specific product/service features customers say they want in a product. An organization must be willing to provide these features in their products to satisfy their customers.

d. Unexpected Requirements

Unspoken features of a product that make the product unique and distinguish it from the competition. Unexpected requirements define product attributes or features that are a pleasant surprise to customers when they first encounter them.

The next step will identify product features by decomposing product(s) into sub-systems and/or sub-assemblies (physical elements). These physical elements are described using metrics and/or product features. Metrics describe the physical characteristics of the component that allow the product to fulfill its function. Metrics will form the basis of the design effort as design teams try to assign parameters to metrics and ensure that the product functionality and requirements has not been violated. It is important to identify the metrics in a manner that will allow the designer the freedom to be creative and the flexibility to choose among different design approaches. The HOQ is then implemented.

Customer requirements identified earlier and categorized according to Kano's model will be arranged into a hierarchy of primary, secondary, and tertiary customer requirements that can be addressed by the engineering staff. Next product metrics that were obtained through product decomposition analysis will be listed. Customer requirements and product features are compared with respect to one another and their relationships are determined using the numerical values shown below:

- Strong relationship: 9
- Medium relationship: 3
- Weak relationship: 1
- No relationship exists: 0

If there is an empty row, this is an indication that a customer need is not being met since that need is not being addressed by any of the product features. In this case, additional product features are identified to satisfy that customer need. An empty column indicates that a particular product feature does not affect any of the customer requirements and may be removed from the house of quality chart. The interrelationships between product features are also identified and recorded. In this case symbols are used to describe the strength of the interrelationships. This correlation matrix identifies which product features support one another and which are in conflict. Conflicting product features could be the result of conflicting customer requirements and represent points at which trade-offs must be made.

Customer competitive assessment is a summary of the top competitive products' features in comparison with the product being developed. Also, customer competitive assessment contains an appraisal of where an organization stands relative to its major competitors in terms of each customer's requirement. Technical competitive assessment is a benchmarking study that compares the competitors' specifications for each of the product features and the proposed specification to either meet or exceed the customer requirements. Customer requirements are then prioritized using:

- *Importance of Customer Needs Rating:* Customer needs are rated according to their importance with respect to each other. The question being answered here is "How important is this feature to the customer?"
- *Target Value:* The quality function deployment team sets values for each product feature depending on whether they (the team) want to

keep the feature unchanged, improve it, or make the feature better than the competition.

- **Scale-Up Factor:** The scale-up factor is the ratio of the target value to the product rating given in the customer competitive assessment. The higher the value, the more effort is needed.
- **Sales Point:** Marketing uses the sales point to determine if they will get any leverage out of any improvement. That is, the sales point answers the question "Taking into consideration the importance of this feature to the customer and the effort needed to make the change, if we change this feature, can marketing get any leverage from it?"
- **Absolute Weight:** The absolute weight is calculated using the following expression:

Importance Rating * Scale-Up Factor * Sales Point

Finally prioritization of product features are performed which includes:

- **The Degree of Technical Difficulty:** A measure of the organization's capability of making a certain product feature.
- **Target Value:** An objective measure that defines values that must be achieved in order to fulfill customer requirements.
- **Absolute Weight:** The absolute weight for the jth product feature is:

$$a_j = \sum_{i=1}^{n} R_{ij} * C_i$$

Where:
a_j = absolute weight for product features ($j = 1,\ldots,m$)
R_{ij} = weight assigned to relationship matrix ($i= 1,\ldots,n, j = 1,\ldots,m$)
C_i = degree of importance of customer requirements ($i= 1,\ldots,n$)
m = number of product features
n = number of customer requirements

5.2 Requirement Analysis

Results of the need analysis step are used to identify the product requirements. The development group begins by preparing a list of *functional objectives* needed to meet the customer's primary needs. Further analysis of customer needs reveals *operational functional requirements* that impose both functional and physical constraints on the design. Secondary customer requirements will be categorized as *general functional requirements*; they are ranked secondary because they will not affect the main function of the product. That is, a product may lack one or more general functional requirement and still be considered as a functional product that meets the intended function. General functional requirements are weighted with respect to their importance. The following provide further discussions on these elements.

i. *Functional Objectives*

Functional objectives are an abstraction of the product function required to satisfy customer needs. They provide information about what the device/product under investigation is supposed to do. Functional objectives can be thought of as the basic operations or transformations that must be performed by the system to satisfy customers' primary needs. Functional objectives are often somewhat general, but they should always employ action phrases such as "to transform", "to support", or "to lift".

ii. *Operational Functional Requirements*

Operational functional requirements are detailed and specific information representing a set of constraints that the design must possess in order to satisfy the product is intended function. Needs analysis will identify the operational conditions and the physical limitations of the product under investigation, which should be translated into operational functional requirements giving quantitative data wherever possible. Operational functional requirements are usually presented in the form of ranges.

iii. *General Functional Requirements*

General functional requirements are the criteria set by the designer, based on the needs analysis, to evaluate the resulting design. General functional requirements are those that satisfy the customers' secondary needs, which could form a critical factor for the customer when comparing different competitive products that accomplish the same function.

iv. *General Functional Requirements Weights*

Several general functional requirements may exist for a product, some are more important than others, therefore different weights should be assigned to different requirements. Customer needs are considered an essential factor in weight assignment. Using a benchmarking study of competitive products could make weight assignment. Alternatively, it could be an input of the design team based on previous knowledge of the importance of such requirements.

5.3 Product Concept Analysis

Product/concept analysis is the decomposition of the product into its basic functional and physical elements. These elements must be capable of achieving the product's functions. Functional elements are defined as the individual operations and transformations that contribute to the overall performance of the product. Physical elements are defined as parts, components, and subassemblies that ultimately implement the product's function.

Product concept analysis consists of product physical decomposition and product functional decomposition. In product physical decomposition, the product is decomposed into its basic physical components which, when assembled together, will accomplish the product function. Physical decomposition should result in the identification of basic components that must be designed or selected to perform the product function. Product functional decomposition describes the product's overall functions and identifies component's functions. Also, the interfaces between functional components are identified.

5.4 Product/Concept Integration

Basic components resulting from the decomposition process should be arranged in modules and integrated into a functional system. The manner by which components are arranged in modules will affect the product design. The resulting modules can be used to structure the development teams needed. Following are steps associated with product integration:

i. Identify System Level Specifications (SLS)

System level specifications are the one-to-one relationship between components with respect to their functional and physical characteristics. Functional characteristics are a result of the operations and transformations that components perform in order to contribute to the overall performance of the product (functional domain). Physical characteristics are a result of the components' arrangements, assemblies, and geometry that implement the product function (physical domain). A general guideline for identifying the relationships are:

a.   *Functional Characteristics*
1. Identify the main function(s), based on the functional decomposition.
2. Identify the required operations and transformations that must be performed in order to achieve the function.
3. Categorize operations and transformations into a hierarchy structure.

b.   *Physical Characteristics*
1. Identify any physical constraints imposed on the product based on the requirement analysis.
2. Identify possible arrangements and/or assemblies of the components, based on previous experiences, previous designs, engineering knowledge, or innovative designs/concepts.
3. Categorize arrangements and assemblies into a hierarchy structure.

Physical and functional characteristics, forming the system level specifications, are arranged into a hierarchy of descriptions that begins by the component at the top level and ends with the detailed descriptions at the bottom level. Bottom level descriptions (detailed descriptions) are used to determine the relationships between components, 1 if the relationship exists and 0 otherwise. This binary relationship between components is arranged in a vector form, "System Level Specifications Vector"(SLSV).

ii. *Identify the Impact of the System Level Specifications on the General Functional Requirements*

System level specifications would affect the general functional requirements in the sense that some specifications may help satisfy some general functional requirements, while other specifications might prevent the implementation of some desired general functional requirements. The impact of the SLS on GFR's are identified which will help in developing products that will meet, up to a satisfactory degree, the general functional requirements stated earlier. The impact will be determined based on,

-1 Negative Impact
0 None
+1 Positive Impact

A negative impact represents an undesired effect on the general functional requirements such as limiting the degree to which the product will meet the general requirement, or preventing the product from implementing the general requirement. While a positive impact represents a desired effect that the SLS will have on the general requirements, such SLS will ensure that the product will satisfy the requirements and result in customer satisfaction. An SLS is said to have no impact if it neither prevents the implementation of the GFR, nor helps satisfying the GFR.

iii. Measure the Similarity Index

The degree of association between components is now measured and used in grouping components into modules. This can be done by incorporating the general functional requirement weights, in addition to the system level specifications vectors and their impacts on the general functional requirements to provide a similarity index between components. The general form of the similarity index is as following:

SLSV (C_1 & C_2)   SLS & FRs   Weights for FRs

$$(S)_{1x1} = \begin{pmatrix} 1 & 0 & . & . & a_n \end{pmatrix}_{1xn} * \begin{pmatrix} 1 & . & . & . & b_{1,m} \\ 0 & . & . & . & . \\ . & . & . & . & . \\ . & . & . & . & . \\ b_{n,1} & . & . & . & b_{n,m} \end{pmatrix}_{nxm} * \begin{pmatrix} 1 \\ 0.9 \\ . \\ . \\ c_{mx1} \end{pmatrix}_{mx1}$$

The similarity indices associated with components are arranged in a component vs. component matrix. Components with high degree of association are grouped together using either an optimization model or any other clustering methods (Kamrani and Salhieh 2002, Singh and Rajamani 1996). These groups or modules can now be developed independently and simultaneously.

6. CASE STUDY

The selected test product for the case study possess moderate complexity to ensure that effort is focused on applying and validating the proposed approach, rather than spent on attempting to understand a complex product. Maintaining moderate complexity will also show the potential for using the proposed approach in designing complex products or systems. A need to design a speed reducer is identified. The speed reducer is part of the power transmission system of a pump. The power is generated by an electric motor that operates at a fairly high speed while the pump must rotate slowly.

Customer needs and requirements are collected and categorized according to Kano's model. Table 5-2 shows a sample of those requirements. The product to be design is decomposed into basic components and the main features/metrics are identified. Table 5-3 is partial list of the identified features/metrics.

Table 5-2. Partial result from Kano Model

No	Customer	Expected	Spoken
1	15 hp must be transmitted from the motor to the	Yes	Yes
2	The speed delivered to the pump must be reduced from (motor speed) to 300	Yes	Yes
3	The motor output shaft and the pump input shaft must	Yes	Yes
4	The space available to install the device is 20"	Yes	Yes
5	The height of the product should not	Yes	Yes
6	The component of the product should have enough space them to allow easy	Yes	No
7	The product should have small size and	Yes	No
8	The product should be capable of operating at temperatures from 0° to 130°	Yes	Yes
9	The product should have a moderate	No	No
10	The product should be safe to	Yes	No
11	The product should be easy to	No	No
12	The product should use standard materials and	No	No

Table 5-3 *Partial component and feature list*

Component	Feature Description	Feature Symbol
Gear	Outside Diameter	G_{OD}
	Inside Diameter	G_{ID}
	Number of teeth	t
Shaft	Center line Height	H1
	Outside Diameter	Sh_{OD}
	Length	L

Customer requirements and product features are used in building the house of quality as shown in Figure 5-7.

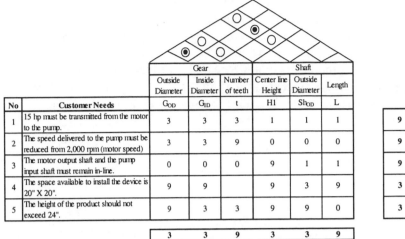

| No | Customer Needs | Gear | | | Shaft | | |
		Outside Diameter G_{OD}	Inside Diameter G_{ID}	Number of teeth t	Center line Height $H1$	Outside Diameter Sh_{OD}	Length L
1	15 hp must be transmitted from the motor to the pump.	3	3	3	1	1	1	9
2	The speed delivered to the pump must be reduced from 2,000 rpm (motor speed)	3	3	9	0	0	0	9
3	The motor output shaft and the pump input shaft must remain in-line.	0	0	0	9	1	1	9
4	The space available to install the device is 20" X 20".	9	9		9	3	9	3
5	The height of the product should not exceed 24".	9	3	3	9	9	0	3
		3	3	9	3	3	9	

Figure 5-7. HOQ for the speed reducer

The prioritized needs from the House of Quality are translated into a set of requirements as shown below:

Operational Functional Requirements:

- The reducer must transmit 15.0 hp
- The input is from an electric motor at a rotational speed of 2000 rpm
- The output delivers the power at a rotational speed range of 290 to 300 rpm
- The input and output shaft must be in-line
- The reducer must be installed on a square surface 20" X 20", with a height of 24"

General Functional Requirements and weights:

- Performance: The degree to which the design meets or exceeds the design objectives
- Compactness: Small size and weight
- Ease of Service: Components should be arranged in a way that it is accessible for maintenance and replacement

All requirements have equal importance and the weights are set to 1. It is assumed that concept generation and concept selection was performed and resulted in the selection of a four-gear speed reducer to accomplish the required function and meet the requirements stated in the previous step.

The overall system is decomposed into four physical sub-systems that includes the speed reducer. Then the speed reducer is decomposed into its basic physical components that are gears, shaft, keys and bearing. The product overall function is conceptualized into an action statement **"To transmit power and reduce speed"**, and represented in a function block diagram (Figure 5-8).

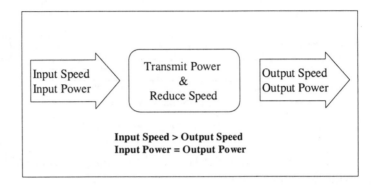

Figure 5-8. Overall Function of the Speed Reducer

The functional objectives are:
- To receive power from an electric motor through a rotating shaft
- To transmit power through machine elements that reduce the rotational speed to a desired value
- To deliver the power at a lower speed to an output shaft that ultimately drives the pump

System level specifications (SLS) are determined based on the functional and physical decomposition. The resulting decomposition is arranged in a hierarchy structure as in Figure 5-9. The one-to-one relationships between components (System Level Specifications) are determined based on the detailed descriptions of the system level specifications. These are located in the bottom level of the hierarchy. The relationship will be assigned 1 if it exists and 0 otherwise. A partial listing of the resulting system level specifications are listed in Table 5-4.

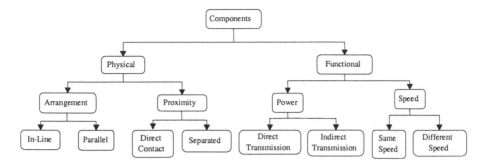

Figure 5-9. System Level Specification Hierarchy Structure

Table 5-4. System Level Specifications

		Components							
		Physical				Functional			
		Arrangement		Proximity		Power		Speed	
Comp. 1	Comp. 2	In-Line	Parallel	Direct Contact	Separated	Direct Transmission	Indirect Transmission	Same Speed	Different Speed
Gear1	Gear 2	0	1	1	0	1	0	0	1
	Gear 3	0	1	0	1	0	1	0	1
	Gear 4	1	0	0	1	0	1	0	1
	Shaft 1	1	0	1	0	1	0	1	0
	Shaft 2	0	1	0	1	0	1	0	1
	Shaft 3	1	0	0	1	0	1	0	1
	Bearing 1	1	0	0	1	0	0	1	0
	Bearing 2	1	0	0	1	0	0	1	0
	Bearing 3	0	1	0	1	0	0	0	1
	Bearing 4	0	1	0	1	0	0	0	1
	Bearing 5	1	0	0	1	0	0	0	1
	Bearing 6	1	0	0	1	0	0	0	1
	Key 1	0	1	1	0	1	0	1	0
	Key 2	0	1	0	1	0	1	0	1
	Key 3	0	1	0	1	0	1	0	1
	Key 4	0	1	0	1	0	1	0	1

From Table 5-4, it can be shown that Gear 1 and Gear 2 have a parallel arrangement and a direct contact. Also, the functional characteristics show that the power is transmitted directly between them and their speed is different. The impact of SLS on GFRs is determined by using the needs analysis and previous knowledge of the system under design. In determining the impact, the designer might ask questions such as: What will happen to the GFR if we did/did not have this specific SLS? or What kind of effect will this SLS have on the GFR? Answers to these questions should identify the impacts. If it is a desired impact, it will be assigned (1). If the impact is undesired or if it will prevent the product from achieving its functions, it will be assigned (-1). And if the impact is not significant or it does not affect the GFR, it will be given a (0). The resulting impacts are shown in Table 5-5.

The similarity index is used to determine the degree of association between the different components. The indexes are arranged in a matrix as shown in Figure 5-10. To group the components p-median model is used to rearrange the similarity matrices into independent modules. The model is as follows:

Notation:
C = Number of components
M = Number of modules
S = Similarity Index

Maximize the sum of all the similarities:

$$\sum_{p=1}^{C} \sum_{q=1}^{C} S_{pq} X_{pq}$$

Where

$$X_{nm} = \begin{cases} 1, \text{ if component n belongs to module m} \\ 0, \text{ otherwise} \end{cases}$$

Subject to:

Each component is assigned to exactly one module:

$$\sum_{q=1}^{C} X_{pq} = 1 \quad , \quad \forall p$$

Components are assigned to a predefined number of modules:

$$\sum_{q=1}^{C} X_{pq} = M$$

Components are assigned to modules that have a median component:

$$X_{pq} \leq X_{qq}, \quad \forall p,q$$

The model is an integer model:

$$X_{pq} = 0/1, \quad \forall p,q$$

The best solution had three independent modules as shown in Figure 5-11.

Table 5-5. Impact of SLS on GFR

Sys. Level Specs	General Functional Requirements		
	Ease of Service	Compactness	Performance
In-Line	1	1	0
Parallel	1	1	0
Direct Contact	0	1	1
Separated	1	-1	-1
Direct Transmission	0	0	1
Indirect Transmission	0	0	0
Same Speed	0	0	0
Different Speed	0	0	1

	Gear1	Gear 2	Gear 3	Gear 4	Shaft 1	Shaft 2	Shaft 3	Bearing	Bearing	Bearing	Bearing	Bearing	Bearing	Key 1	Key 2	Key 3	Key 4
Gear1		5	1	1	6	1	1	2	2	1	1	1	1	6	1	1	1
Gear 2	5		2	1	1	6	1	1	1	2	2	1	1	1	6	3	2
Gear 3	1	2		2	1	6	1	1	1	2	2	1	1	1	1	6	1
Gear 4	1	1	2		1	1	6	1	1	1	1	5	5	1	1	1	6
Shaft 1	6	1	1	1		1	1	5	5	1	1	1	1	6	1	1	1
Shaft 2	1	6	6	1	1		1	1	1	5	5	1	1	1	6	6	1
Shaft 3	1	1	1	6	1	1		1	1	1	1	5	5	1	1	1	6
Bearing 1	2	1	1	1	5	1	1		2	1	1	1	1	2	1	1	1
Bearing 2	2	1	1	1	5	1	1	2		1	1	1	1	2	1	1	1
Bearing 3	1	2	2	1	1	5	1	1	1		2	1	1	1	2	2	1
Bearing 4	1	2	2	1	1	5	1	1	1	2		1	1	1	2	2	1
Bearing 5	1	1	1	5	1	1	5	1	1	1	1		2	1	1	1	2
Bearing 6	1	1	1	5	1	1	5	1	1	1	1	2		1	1	1	2
Key 1	6	1	1	1	6	1	1	2	2	1	1	1	1		1	1	1
Key 2	1	6	1	1	1	6	1	1	1	2	2	1	1	1		2	1
Key 3	1	3	6	1	1	6	1	1	1	2	2	1	1	1	2		1
Key 4	1	2	1	6	1	1	6	1	1	1	1	2	2	1	1	1	

Figure 5-10. Similarity Matrix

	Gear1	Shaft1	Bearing1	Bearing2	Key1	Gear2	Gear3	Shaft2	Bearing3	Bearing4	Key2	Key3	Gear4	Shaft3	Bearing5	Bearing6	Key4
Gear 1		6	2	2	6	5											
Shaft 1	5		5	5	6												
Bearing 1	2	5		2	2												
Bearing 2	2	5	2		2												
Key 1	6	6	2	2													
Gear 2	5						2	6	2	2	6	3					
Gear 3						2		6	2	2	1	6					
Shaft 2						6	6		5	5	6	6					
Bearing 3						2	2	5		2	2	2					
Bearing 4						2	2	5	2		2	2					
Key 2						6	1	6	2	2		2					
Key 3						3	6	6	2	2	2						
Gear 4														6	5	5	6
Shaft 3													6		5	5	6
Bearing 5													5	5		2	2
Bearing 6													5	5	2		2
Key 4													6	6	2	2	

Figure 5-11. Independent Modules

These modules could be designed simultaneously. The solution also illustrates the interactions that exist between the independent modules. These interactions are represented by the similarity between different components that belong to independent modules. The association between components belonging to different modules could be used to identify the type of the interactions and thus identify the information required to explain this interaction.

7. Conclusion

Modularity provides a high degree of independence among elements. Through axiomatic design and application of modularity, the product development cycle time is greatly reduced. This is achieved through the design and development of the individual modules. An uncoupled design allows for ease of design upgrade and improvement. Through this independence, the functional capability of the product can be redesigned and improved. The proposed methodology incorporated the VOC into its process. QFD concept provides product designers with a structured planning approach to identify customer requirements, and Design for Modularity develops independent and standard components, which could migrate from one product design to another. Combining both concepts resulted in a systematic approach that can be used to identify components that can be developed in parallel while meeting customer requirements. The combined approach introduces a parallel development cycle that will reduce the development time, and allow cross-functional team participation, furthermore the combined approach will allow organizations to respond effectively to customer need.

REFERENCES

1. Besterfield D.H., Besterfield-Michna C., Besterfield G.H., Besterfield-Sacre M. Total Quality Management, Prentice-Hall Inc., 1995.
2. Finkelman D.P., Goland A.R. How Not to Satisfy Your Customers. McKinsey Quarterly 1990; Winter: 2–12.
3. Kamrani A., Salhieh S. Product Design for Modularity, 2nd Edition. Kluwer Academic Publishers, 2002.
4. Kano N., Seraku N., Takahashi F., Tsuji S. Attractive Quality and Must-Be Quality. Hinshitsu: The Journal of the Japanese Society for Quality Control 1984; April: 39–48.
5. Matzler K., Hinterhuber H. H. How to Make Product Development Projects More Successful by Integrating Kano's Model of Customer Satisfaction into Quality Function Deployment. Technovation 1998; 18/1: 25–38.
6. Pahl G., Beitz W. Engineering Design: A Systematic Approach. London; New York: Springer, 1996.
7. Pugh S. Total Design: Integrated Methods for Successful Product Engineering. Wokingham, England; Reading, Massachusetts: Addison-Wesley Publishing Co., 1991.
8. Salhieh S., Kamrani A. Macro Level Product Development Using Design for Modularity. Robotics and Computer Integrated Manufacturing Journal 1999; 15: 319-329.
9. Shillito, L.M. Advanced QFD: Linking Technology to Market and Company Needs New York: John Wiley & Sons, 1994.
10. Shirly G.V., Modular Design and the Economics of Design for Manufacturing. Integrating Design and Manufacturing for Competitive Advantage, G. Susman, editor, Oxford University Press, 1992.
11. Singh N., Rajamani D. Cellular Manufacturing Systems: Design, Planning, and Control. Chapman & Hall, 1996.
12. Suh N.P. Design and Operation of Large Systems. Journal of Manufacturing Systems 1995: 14/3: 203-213.
13. Sullivan L.P. Quality Function Deployment. American Society for Quality Control. Quality Progress 1986; June.
14. Ulrich K.T., Eppinger S.D. Product Design and Development, 2nd Edition. New York: McGraw-Hill, 2000.
15. Ulrich K., Tung K. Fundamentals of Product Modularity. Issues in Design, Manufacture and Integration 1991; DE-Vol. 39, ASME.

CHAPTER 6

INTEGRATED PRODUCT DESIGN AND DEVELOPMENT IN COLLABORATIVE ENVIRONMENT

Ali K. Kamrani[1], Mahesh Kanawade[2]

[1]*University of Houston*
[2]*University of Michigan - Dearborn*

Abstract   Increased global competition has forced organizations to enhance product variety and shorten time-to-market. Also, vendors' early participation in the design process is critical for improving product quality and drastically reducing development cycle time. The collaborative approach abridges these needs. This chapter presents an integrated framework for product design and development in distributed and collaborative environment. The proposed methodology emphasizes integration of software tools and resources.

Keywords: Collaborative engineering, integrated design, catalog-based design, collaborative CAD modeling.

1. INTRODUCTION

During the past decade, global manufacturing competition has increased significantly. Consequently, the manufacturing industry in the United States has been undergoing some fundamental changes, including a move to low-cost, high-quality systems and a shift in focus from large business customers to a diffused commodity market for all sizes and types of customers. Obstacles include shortened product life cycles, high-quality products, highly diversified and global markets, and unexpected changes of technologies and customer needs. As a result, companies are heading toward vendor-based manufacturing, i.e., manufacturers are trying to get most of the work done by vendors so as to minimize the time-to-market.

The focus on a customer-driven market, coupled with increased competition, requires fast updating of designs, flexibility in manufacturing systems, and responsiveness in production schedules. Two of the more important elements in today's changing environment are increased product sophistication and variation. To remain competitive, manufacturers must minimize total costs while being quick to develop and market new products. This involves integrating many diverse functional areas of an organization into a process of creating a better design when viewed across the entire product life cycle. One of the consequences of the demand for this integration of resources has been the necessity for teams of engineers, often from several areas and geographical locations, to work together over networks, supported by information and computer services. The major problem here is the integration of various types of data involved in an enterprise; how to deal with this diversified data and information.

2. PRODUCT DESIGN AND DEVELOPMENT PROCESS

Product development is the process of creating a new product to be sold by a business or enterprise to its customers. Design refers to those activities involved in creating the styling, look, and feel of the product; deciding on the product's mechanical architecture; selecting materials and processes; and engineering various components necessary to make the product work. Development refers collectively to the entire process of identifying a market opportunity; creating a product to appeal to the identified market; and finally testing, modifying, and refining the product until it is ready for production.

The task of developing products is difficult, time-consuming, and costly. Noteworthy products are not simply designed, but instead they evolve over time through countless hours of research; analysis; design studies; engineering and prototyping efforts; and finally testing, modifying, and re-testing until the design has been perfected.

The impulse for a new product normally comes from a perceived market opportunity or from the development of a new technology. Consequently, new products are broadly categorized as either market-pull products or technology-push products. With a market-pull product, the marketing department of the firm first determines that sales could be increased if a new product were designed to appeal to a particular segment of its customers. Engineering is then asked to determine the technical feasibility of the new product idea. This interaction is reversed with a technology-push product. When a technical breakthrough opens the way for a new product, marketing then attempts to determine the idea's prospects in the marketplace. In many cases, the technology itself may not actually point to a particular product but instead to new capabilities and benefits that could be packaged in a variety of ways to create a number of different products. With either scenario, manufacturing is responsible for estimating the cost of building the prospective new product, and their estimations are used to project a selling price and estimate the potential profit for the firm. If the decision has been taken to outsource some of the components in the final product, the vendors come into direct consideration. The vendors become a part of the design team, as they will be contributing towards the final product. Hence, it is very important to consider the vendors' involvement in the design process beginning from the initial stages of the design and development of the product.

2.1   Integrated Product Development (IPD)

Very few products are developed by an individual working alone. It is unlikely that an individual will have all the necessary skills in marketing, industrial design, mechanical and electronic engineering, manufacturing processes and materials, tool-making, packaging design, graphic art, project management, etc. Development is normally done by a design team as an *integrated* approach. The team leader draws on talent in a variety of disciplines, often from both outside and inside of the organization. As a general rule, the cost of a development effort is a factor of the number of people involved and the time required for fostering the initial concept into a fully refined product.

Integrated product development (IPD) practices are recognized as critical to the development of competitive products in today's fast-paced global economy. Product development teams, particularly when team members are collocated, are a critical element of IPD practices to facilitate early involvement and parallel design of products and their processes. As a firm grows larger and products become more complex, hierarchical organizations are established to handle the increasingly large organization size, the technical complexity, and the specialization that evolves to master this complexity. This firm growth also results in the geographic dispersion of people and functional departments. These factors inhibit many of the informal relationships that previously provided effective communication and coordination between functions. Functional departments tend to focus inwardly on functional objectives. This is often described as the functional bin. A hierarchical organization structure with enterprise activities directed by functional managers becomes incapable of coordinating the many cross-functional activities required to support product development as the enterprise moves toward parallel design of product and process and a focus on time-to-market. *Product development teams* (PDTs) are a way to address this complexity by organizing the necessary skills and resources on a team basis to support product and process development in a highly interactive, parallel collaborative manner.

2.1.1   Principles of Integrated Product Development

Some of the basic principles and guidelines for *integrated product development* are listed below:

1. *Understand Customer Needs and Manage Requirements:*
   Customer involvement increases the probability of the product meeting those needs and being successful in the market. Once customer requirements are defined, track and tightly manage those requirements and minimize creeping elegance that will stretch out development.

2. *Plan and Manage Product Development:*
   Integrate product development with the business strategy and business plans. Determine the impact of time-to-market on product development and consider time and quality as a source of competitive advantage.

3. *Use Product Development Teams:*
   Early involvement of all the related departmental personnel in product development provides a multi-functional perspective and facilitates the parallel design of product and process, reducing design iterations and production problems. Collocation improves communication and coordination among team members.

4. *Involve Suppliers and Subcontractors Early:*
   Suppliers know their product technology, product application, and process constraints best. Utilize this expertise during product development and optimize product designs.

5. *Integrate CAD/CAM and CAE Tools:*
   Integrated CAD/CAM/CAE tools working with a common digital product model facilitate capture, analysis, and refinement of product

and process design data in a more timely manner. Feature-based solids modeling, parametric modeling, and electronic design frameworks facilitate the downstream interpretation, analysis, and use of this product data.

6. *Simulate Product Performance and Manufacturing Processes Electronically:*
   Solids modeling with variation analysis and interference checking allow for electronic mock-ups. Analysis and simulation tools such as FEA, thermal analysis, NC verification, and software simulation can be used to develop and refine both product and process design inexpensively.

7. *Improve the Design Process Continuously:*
   Re-engineer the design process and eliminate non-value-added activities. Continued integration of technical tools, design activities, and formal methodologies will improve the design process.

3. COLLABORATIVE ENGINEERING APPROACH

Engineering paradigms like participatory design, concurrent design and TQM all focus on teamwork. Participatory design supports cooperation between users and system designers. Concurrent engineering especially focuses on cooperation between design and production. TQM requires cooperation between all departments of an enterprise. Collaborative engineering is an innovative method for product development that integrates widely distributed engineers for virtual collaboration. The reasons for geographically widely dispersed teams as depicted in Figure 6-1, are various, e.g., locality of certain resources and competence or perhaps different production costs. Computer modeling is used in the whole engineering design process, resulting in virtual prototypes. High-edge technology is required to assure a real-time, interactive engineering process. This includes high-performance workstations with advanced visualization and modeling software, high-speed networks for data transfer, compatible data exchange media, and appropriate standards including those for product data representation.

The need for integration continues when the design enters the preliminary and detail design phases. In the virtual, integrated, concurrent design environment, designers interact by sharing information and reaching agreements. By considering proper integration and interaction from the beginning, problems with the final integration of activities will be significantly reduced. In this context, an integrated system is desired to reduce the overall design cycle time by eliminating repetitive calculations to obtain optimum results.

3.1 Role of Collaboration in Integrated Product Development

Collaborative engineering is considered to be important for any engineering firm wishing to survive in the present market. The growing complexity of products and the design process has elevated the importance of efficient product design and development. The pressure for reducing the cycle time of product development, engineering and manufacturing has made it necessary to improve the integration of these functions and to mange the interfaces. Collaboration is one of the central requirements for engineering today. The shifting from the traditional design and manufacturing paradigm to a new, virtual and agile model is generally observed. The traditional model is one with very limited information sharing, static organizational structure, and almost no cooperation among competitors.

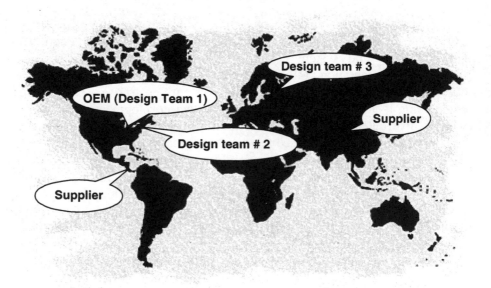

Figure 6-1. Design scenario – Participants in design team are distributed geographically. Each presents own viewpoint and software tool, but unerstanding of the process as a complete product is needed

3.2 Product Development Teams and Collocation

Product development teams are formed with personnel from various functional departments to support different stages of the development process, including production and services as shown in Figure 6-2. This early involvement will result in a complete understanding of all the requirements and a consensus approach to the design of both the product and its manufacturing and support processes. Product development teams promote open discussion and innovative thinking resulting in superior products, more efficient processes, and, ultimately, a more satisfied customer. The focus of the team will be to satisfy the external customer's product and support requirements as well as the internal customer (functional department) requirements related to factors such as producibility, cost, supportability, testability, etc.

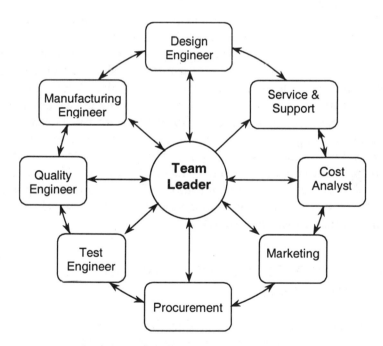

Figure 6- 2. Collaborative Team Composition

Although PDTs require more resources early in the development cycle, the result will not only be superior designs, but reduced resources over the life cycle of development, production, and support through reduced design iterations. The team approach will lead to greater commitment to the design and will result in a smoother transition to production.

3.3 Effectiveness of PDT

A key factor in the effectiveness of the PDT is the opportunity for regular interaction among team members and working together as a true team. If a team only meets periodically much as a committee would, interaction, working relationships, collaboration, and effectiveness are limited. The majority of potential communication among team members regarding a product development effort is informal. The physical proximity of an "expert" in another discipline will trigger asking a question and seeking or sharing information. Collaboration not only facilitates this communication but improves the nature of working relationships and leads to a more streamlined development process. This improved informal communication and coordination accelerates development activities and truly enables a parallel mode of development.

The important stage of collaboration is breaking down barriers between departments that results in throwing designs over the wall. As personnel have an opportunity to interact and develop relationships, issues and questions can more easily be resolved. Personnel learn of others' expertise, and there is a greater opportunity to take advantage of "hidden knowledge." As the organization moves toward implementing PDTs, the closer proximity of the functional departments makes it easier for the team members to work together and coordinate activities. As individuals are working on a day-to-day basis in close proximity, they have an opportunity to develop a close working relationship that improves overall team dynamics. This enhances the frequency and quality of communication. There is greater opportunity for feedback and discussion. Team members can respond more rapidly to issues and initiate process tasks more quickly.

4. PRODUCT DESIGN IN COLLABORATIVE ENVIRONMENT

As described in section 3, product design requires team efforts and cannot be initiated and completed by an individual single-handedly. Currently, industry recognizes the need for overlapping of tasks to reduce the design cycle time. Any product development activity starts with the development of a product model. This product model is utilized for design analysis manufacturing feasibility, etc. In a *collaborative environment*, the product model representation should capture all the geometric and technological information that will be required for further analysis. The visual representation is usually developed in CAD packages, and the technological data such as surface roughness, material, etc. should be closely associated with it. This will help the designers to easily access and modify the information at every stage of the development process.

In a traditional paradigm, engineering occurs in a sequential manner. Design must occur largely before manufacturing; testing must occur after manufacturing has started but before full-scale production is undertaken. However, *collaborative engineering* is based on the observation that there is no necessary condition forcing that expertise to occur only at the required stage. In most of the cases, expertise from several fields is absolutely essential at every stage of product development. There is a need to provide a variety of expertise concurrently at each stage of a sequential process. These varieties of expertise are grouped in different teams. Each team is responsible for its respective contribution throughout the process. One should not interpret the term *expert* as a consultant. These are the active members throughout the development process. With these concurrent experts, we can begin to parallelise tasks. Doing engineering tasks in parallel tends to shorten lead-times, which improves the overall efficiency of the product development process.

For reduction in product development cycles, it is necessary to capture complete product model information that can be utilized by product design modules for further processing. Also, the product models should be accessible to geographically dispersed product developers (i.e., designers and process planners) as well as remotely located product design modules. This necessitates a *collaborative integrated framework* that can meet the need for this increased agility in production organizations.

4.1 Collaborative Marketplace

A major problem in achieving effective and timely product designs and product innovations has been the long design and product development process. The long design development process can be attributed to the need for many design iterations and problems typically solved with meetings and many phone calls. Many times, work is accomplished in informal collaboration, where the emphasis is on exploration of the ideas, compared to formal collaboration, which mostly consists of confirming designs that are brought to the meetings. However, traditional collaboration is geographically limited. Colleagues are not easily able to collaborate and exchange their ideas if they are situated in different locations.

The collaborative marketplace has been evolving over the last 15 years, delivering technologies that enable coordination and information sharing, virtual meetings, and more recently virtual collocation. The promise of these technologies is to improve organizational ability to collaborate, coordinate, and share information in order to facilitate inter- and intra-organizational teamwork (Khalid 2001). While these new collaborative media promise to reduce cost and time of information exchange, they have implications for the collaborative design processes. The need to bring together the process of collaboration and computing has led to the development of various collaborative systems. A major purpose of using collaborative systems is to have meaningful interactions with other people. Such richness of interaction can be achieved when barriers of space, time, and media/document formats are overcome when interacting with others. This includes the ability to talk, see, write, and draw in both synchronous and asynchronous manners; access to relevant information; archiving of interactions for future review; and debate of issues on a global basis.

Communication between different members of a design team is a notoriously difficult problem, especially at the early stages of the design process. A recurrent problem in many design domains is communication between members of the design team involved in the different stages of concept creation, embodiment and detail design, fabrication, and production. It is frequently the case that misunderstandings arise such that incorrect and/or incomplete specifications are passed from one team to another and that errors or inconsistencies in the design are discovered at late stages of the process. Sometimes design problems are resolved by production or technical staff, but further re-corrections may be too late or too costly to attempt. A response to this problem has been the development of broader, more integrated teams and the introduction of the "collaborative design process." The communication problem in design teams has also been addressed by the development of computer-based models of the elements being designed. Such models allow simultaneous access by different members of the team, and the element design evolves through collaborative input and evaluation from all of the team.

In an organization for which the design team is geographically dispersed, it is necessary to provide a medium to inter-link its data or work. Nowadays with the advent of World Wide Web applications it is possible to transfer the data or link the design team members. Still, there are numerous problems involved in integrating different software with different operating platforms.

4.2 Collaborative Approach

Collaborative engineering (CE) is the systematic approach to the integrated, concurrent design of products and related processes, including manufacturing, product service, and support. This approach is intended to cause the developers to consider all elements of the product life cycle from conception through disposal, including quality, cost, schedule, and user requirements. The objective of Collaborative Engineering is to reduce the system/product development cycle time through a better integration of resources, activities, and processes.

The basic principle of CE is the integration of methodologies, processes, human beings, tools, and methods to support product development. CE is multidisciplinary in that it includes aspects from knowledge-based systems, hypermedia, database management systems, and CAD/CAM. Collaborative engineering involves the interaction of a diverse group of individuals who may be scattered over a wide geographic range. To enable effective and complete communication among them, there are certain technological concepts that must also become organized into concurrent layers. Distributed information sharing and cooperative work are important techniques to provide a basis for it.

For a CE approach to be effective, there must exist a strong level of communication between developers and end-users. In the context of CE, a customer is both internal and external to the development process. Each member of the CE staff is an internal customer for intermediate products during development. CE advocates an integrated, parallel approach to design. By paying attention to all aspects of the design at each phase, errors are detected prior to being implemented in the product. The integrated design process must include a strong information sharing system, an iterative process of redesigns and modifications, trade-off analysis for design optimization, and documentation of all parts of the design. Integration of computerized systems largely enhances the benefits of CE with automatic knowledge capture during the development and lifetime management of a product and with automatic exchange of that knowledge among different computer systems.

5. RELATED WORK IN COLLABORATIVE PRODUCT DESIGN

Many researchers are working on developing the technologies or infrastructure to support the distributed design environment. Some are working on providing a platform for sharing or coordinating the product information via the World Wide Web. Others are developing the framework that enables the designers to build integrated models and to collaborate by exchanging services (Pahng et al 1998). The major projects under development are DOME (Senin et al 1997) and SHARE (Cutkosky et al 1993).

5.1 DOME (Distributed Object Modeling and Evaluation)

The DOME—Distributed Object Modeling and Evaluation—framework allows the designer to define mathematical models in a collaborative design environment using a set of interconnected modules (Pahng et al 1998). The DOME framework is based on decomposing the multidisciplinary problems into modular sub-problems. This decomposition divides the overall complexity of the problem and distributes the knowledge and responsibility amongst the designers. The DOME framework enables the designer to define and develop mathematical models and link them to form a large system model (Senin et al 1997). Each module summarizes a model describing related aspects of the design in the form of variables and their relations. A DOME module structure is shown in Figure 6-3.

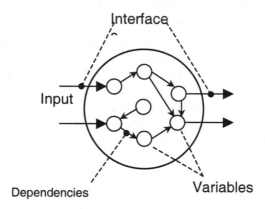

Figure 6-3. DOME Module. The variables form a module. Dependencies represent interaction among the variables.

These modules interact with each other through standard Internet communication protocols, exchanging the information and reacting to each other's changes. A MoDeL—Model Definition Language (Borland 1997)—has been developed for defining problem models in the DOME framework. An optimization engine based on genetic algorithms has been developed and integrated with the framework (Senin and Wallace 1997). Design solutions can be assessed and compared with each other using a goal-based evaluation tool embedded in the framework. The results and the process can be visualized or conveyed to all participants depending on the level of involvement. The structure of DOME is based on distributed object technology and uses CORBA as a communication protocol.

5.2 SHARE Project

The SHARE project explores and applies information technologies to assist the design team to gather, organize, re-access, and communicate both formal and informal design information. It is called a "methodology and environment for collaborative product development" (Cutkosky et al 1993). SHARE project is based on establishing a "shared understanding" of the design and design process among the design teams. The term *understanding* refers to the information, knowledge, viewpoint, decisions, and suggestions of individuals in the design team to share with each other. SHARE advocates computer use in all the design-communication-documentation activities. SHARE methodology is a mixed network-oriented environment that enables the designers to participate in a distributed team using their own tools and databases. It provides templates and commercial software to assist the process of structuring, organizing, and sharing information. It has the following characteristics:

- Online availability of: notebooks, handbooks, design libraries, and related documents
- Collaboration services: electronic mail, video conferencing
- Online catalogs: pricing, shipping, and ordering
- Specialized services: for analysis, simulation, planning, sharing engineering knowledge
- PDM: distributed product data management service to notify concerned personnel of the changes
- Integrated infrastructure: to enable heterogeneous design tools and databases to work transparently

5.3. Limitations of available technologies

An integrative framework that enables designers to rapidly construct performance models of complex problems can provide both design insight and a tool to evaluate, optimize, and select better alternatives. Interaction identification between the elements at each level of design is critical since different patterns of relationships may develop different results as they are integrated. The processes can be used as templates by the designer, which can assist him/her in evaluating and optimizing the design alternatives through proper integration and analysis.

6. INTEGRATED PRODUCT DESIGN AND DEVELOPMENT IN COLLABORATIVE ENVIRONMENT

This system is referenced from the ongoing project at University of Michigan-Dearborn, under Professor Ali Kamrani.

6.1   Overview of the Proposed System

The design teams and vendors operate in different environments. Hence, it is necessary to take into account the vendors and the design teams as a whole system. The design problems are decomposed into models such as physical components, parametric models, or analysis procedures. The important aspect of the proposed framework is an integration of these models used during the design process in the collaborative environment. Thus, the proposed collaborative framework allows the integrated model to be revised with any changes made by individuals in the models involved. The individual does not have to analyze the scenario repeatedly for every change in design variables and validate it for each instance. This framework allows the designer to collaborate with the vendors and other team members to speed up and optimize the design process considering the relationships within these models. *Parametric modeling* is introduced to take advantage of its quicker response, accuracy, consistency, documentation, etc.

The proposed framework is shown in the Figure 6-4. It is composed of four basic phases:

(i)   Analysis Tool
(ii)   Collaborative Environment
(iii)   Optimization Module
(iv)   CAD Modeling

The analysis phase consists of the user interface and the analysis tool. With the design variables entered by the user, a design problem is defined

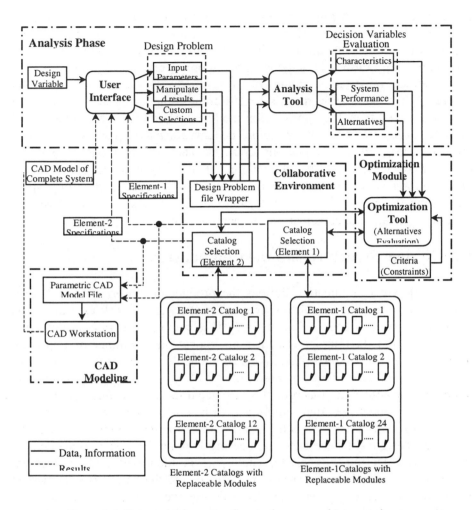

Figure 6-4. Data and information flow in the proposed integrated system

with the ModelCenter wrapper file. The design problem is then analyzed with the analysis tool to generate the decision variables and alternatives. These alternatives are then further analyzed along with the optimization criteria. The vendors contribute to the design process by providing detailed specifications for their components in the form of databases. These databases are converted into the *replaceable modules*, which all together contribute to a *catalog*.

The optimization tool maps these replaceable modules from the catalogs for every instance and places them in the current design alternative until the suitable match is found. The optimization module selects the components from the catalogs and sends them to the user interface as *results*. This gives the user detailed specifications for the product and its elements. These specifications are the parameters obtained from the catalogs, which gives the optimum configuration of components based on the given design variables.

Another important module of the framework is *parametric CAD modeling*. The parametric models for different components are created. The results obtained from the optimization module are used to create the 3D solid models of each element in the system. The term *parametric* means "controlled by parameters," that is, equations or rules. These CAD models assist the designer in visualizing the interaction of the components for a given configuration. An automated design dramatically reduces the time spent generating the results for several alternatives. Also, it serves as a basis for generating detailed documentation for manufacturing.

6.2   System Structure and the Components

The product architecture defines the product in the primary functional systems and subsystems and how they interact so as to work as a unit. The architecture of the product, i.e., the way it is decomposed into systems and subsystems and their integration, has an impact on a number of important attributes such as standardization of components, modularity, options for future changes, ease of manufacture, etc. With outsourced components, the supplier may contribute much of the associated design and engineering. The early involvement of the suppliers, with outsourced components and engineering, can affect the quality and speed of the development process. In the detail design, the necessary engineering is done for every component of the product. During this phase, each part is identified and engineered. Tolerances, materials, and finishes are defined, and the design is documented with drawings or computer files. Considering the recent trend among the manufacturers and developers to generate three-dimensional solid models of the components, solid models of components and their assemblies are created. Three-dimensional computer models form the core of today's rapid prototyping and rapid manufacturing technologies. Once the detailed specification of components is prepared, prototype components can be rapidly built on computerized machines such as CNC mills, fused deposition modeling devices, or stereo-lithography systems and other prototyping techniques.

A collaborative approach for design of a single-stage speed reducer with a pair of spur gears is explained using the proposed framework. The design problem is composed of analysis of geometric details, overall performance, and optimization (compact design) and their interdependency. The design decomposition and different elements of the system are shown in Figure 6-5.

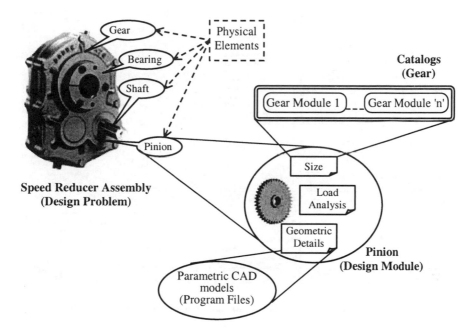

Figure 6-5. Product Decomposition

6.2.1   Collaborative Environment

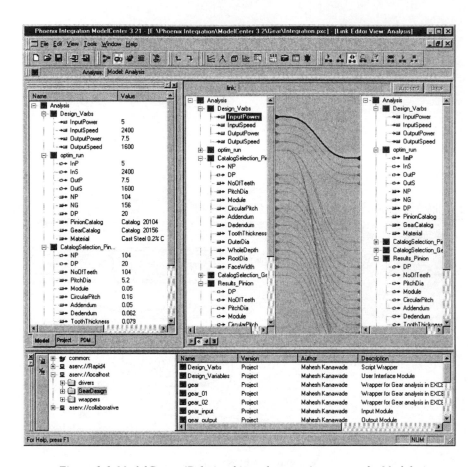

Figure 6-6. ModelCenter (Relationship and connection among the Modules)

The collaborative environment is the integration of all the *modules* and the *catalogs* in the system. The ModelCenter is used as a service protocol that connects different modules and catalogs, keeping the corresponding relationships as shown in Figure 6-6. The file wrappers are created to link the input data file (user interface) and the analysis tool. Once the analysis tool calculates the preliminary parameters, the file wrapper updates the values for the optimization module. The results of optimization are returned to the user interface where the user can comprehend these for further development. At the same time, this set of results is also sent to the CAD modeling module. Solid models are automatically generated for different components of the product.

6.2.2   Analysis Phase

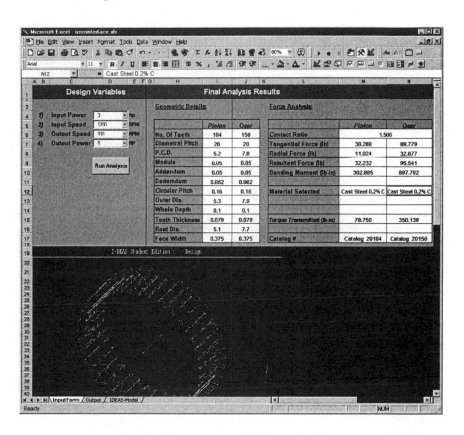

Figure 6-7. User interface with design variables, results obtained, and the solid model for one of the components

The analysis phase as described earlier is composed of a user interface as shown in Figure 6-7. The user interface gives the designer the choice to select or manipulate the design variables and also to comprehend the results after analysis. The provision is made to manipulate the results for specific condition(s) (such as selecting different material, gear, etc.). Another important part of this phase is the analysis tool. This would be an application on the standalone station of the design and validation engineering department. The analysis tool is introduced to set the decision variables and their evaluation. This application anatomizes the design problem defined by the user and gives the feedback in terms of the load characteristics and the performance requirements for different elements in the system. The analysis tool is linked with the user interface through the ModelCenter file wrapper utility.

6.2.3   Optimization Phase

The preliminary results obtained from the analysis tool are used as a base for optimization. These results give theoretical values for the parameters that are acquired from empirical relations. The optimization model is developed to get the configuration with the smallest possible size. The relationships are imposed between the design variables from the design problem and the constraints from the modules. The optimization tool runs an iterative procedure each time the modules from the catalogs are called and put into the current design instance. This procedure runs until the design objective (in this case to have a compact size gearbox) is satisfied. In some cases, multiple results are possible. If this is the case, different results are considered as alternatives and they are re-evaluated with the optimization tool.

6.2.4   Parametric CAD modeling

The results obtained from the optimization are used to create the CAD models. *Parametric models* are created for each component in the system. Once the optimization module generates the results, they are conveyed to the parametric models in the CAD system. Upon getting these results, the CAD tool is initialized and the models are created automatically. The solid models of the components are generated and can be displayed at the user interface in picture format. This gives the designer a chance to visualize the different alternatives and the optimum configuration of the component. These CAD models as a repository can further be used for FEA analysis, NC code generation, manufacturing documentation, reuse for new products, and several other applications.

6.3 Advantages of the proposed system

The framework provides the means of integrating software tools that enables the designers to foresee the overall product and enterprise fulfillment during the development phases. It will reduce the time required for repetitive analysis for different alternatives. Thus, the designer can evaluate more alternatives and obtain the optimal solution. This integrated system allows the designers to participate in the design process irrespective of geographical location. The developed system provides the capability for design of templates for catalog-based design. The vendors can participate in the development process with their items as *catalogs*. The *optimization* phase allows the designers to evaluate different alternatives and tradeoffs.

Some of the advantages of the system are:

- The system allows the integration of the design process with the sub-systems at different locations and the optimization of different elements.
- Rapid analysis of possible alternatives with optimally balanced requirements.
- Flexibility of change, extension, revision, and reuse of models.
- Provides the data for the generation of manufacturing documentation. The developed framework is only the design part of the complete collaborative product design and life cycle assessment.

Future work will include the use of generated models in the further cycle. It may include tasks such as automatically generating process plans, shop drawings, bills of material, machine-control (CNC) code, shop assignments, and documentation for other pieces of the manufacturing process.

7. CONCLUSION

Today's manufacturers encounter various difficulties involved in the product development process that must be overcome for international competitiveness. The obstacles include shortened product life cycles, high-quality products, highly diversified and global markets, and unexpected changes of technologies and customer needs. Any delays in development, and you run the risk of losing revenue to your competition. Also, companies are heading toward vendor-based manufacturing, i.e., the manufacturers are trying to get most of the work done by the vendors so as to minimize the time-to-market. Hence, it is essential to utilize a computer-aided system in designing, manufacturing, testing, and distributing the products to minimize time-to-market.

For the integration of information at every stage of product development, collaboration technology for cooperative work is needed. As the assistant of the design and development of new products, integrated design technology plays a very important role. The framework described in section 6 confirms design assumptions and predicts product performance in the early stages of the design process. This results in a faster product development cycle—with lower associated costs—achieved by eliminating the need to constantly build, test, and redesign.

REFERENCES

1. Borland N. DOME - MoDeL Language Reference. MIT, Cambridge, MA; 1997.
2. Cutkosky M., Toye G., Leifer L., Tenenbaum J. M., Glicksman J. SHARE: A Methodology and Environment for Collaborative Product Development. Post-Proceedings of IEEE Infrastructure for Collaborative Enterprise; 1993.
3. Khalid H. M. Towards Affective Collaborative Design. Proceedings of HCI International; 2001; Mahwah. New Jersey.
4. Pahng F., Senin N., Wallace D.R. Web-Based Collaborative Design Modeling and Decision Support. ASME-DETC 98; 1998.
5. Senin N., Borland N., Wallave D.R. Distributed modeling of Product design problems in a Collaborative design Environment. CIRP International Design seminar Proceedings: Multimedia Technologies for Collaborative Design and Manufacturing; 1997.
6. Senin N., Wallace D.R. A framework for mixed parametric and catalog based pruct design, MIT, CADLab; 1997.

CHAPTER 7

COLLABORATIVE MANUFACTURING FOR MASS CUSTOMIZATION

Mary J. Meixell[1], S. David Wu[2]

[1] *George Mason University*
[2] *Lehigh University*

Abstract: This chapter examines the role of supply chain collaboration in a manufacturing environment where products are mass customized. Specifically, we look at how the structure of the supply chain influences performance where decisions between tiers are coordinated and when product differentiation is postponed through product and process design. We submit that the resulting component commonality has a beneficial effect on the bullwhip effect and on overall performance, and investigate the planning conditions under which these benefits are realized.

Key words: Production planning, bullwhip effect, supply chain management.

1. INTRODUCTION

In recent years, advances in information technology have provided new opportunities for firms who seek to collaborate with their suppliers and customers. Emerging technological infrastructures in both in-house and in the Internet platforms have changed dramatically the economics of sharing information and integrating processes between trading partners. Rapid advances in enterprise information systems have made it possible for large and small firms alike to utilize information that was once too costly to incorporate into daily decision making. Perhaps the greatest advances are in the Internet-enabled software applications that provide supply chain partners with shared information, or that solve problems using decision technologies that had been available but difficult to apply because of laborious data collection (Sodhi 2001). With enabling infrastructure, enterprise and inter-enterprise data systems, and sophisticated application software, supply chain managers are poised to achieve enhanced levels of performance based on improved coordination with their trading partners.

Well-designed supply chain processes are structured to respond to specific customer priorities through a well thought out and implemented supply chain strategy. Mass customization is a particular strategy in supply chain management that can be thought of as the mass production of individually customized goods and services. Typically, both delivery promptness and price are important to the customer in this environment, motivating manufacturers to prioritize both delivery fulfillment time as well as cost. No doubt it is easier to achieve high performing supply chains when finished products are relatively few in number. But where high variety is important to the corporate strategy, the operation of a firm needs to overcome the obstacles and achieve the dual objectives of reasonable price and delivery promptness.

Postponement has been employed by many firms to achieve both low cost and fast delivery times for mass customized products. Postponement delays the point at which variety is introduced in the production process, so that individual items in production are kept in a generic form as late as possible on the production line. For the end item manufacturer, postponing differentiation is a recognized approach to efficient and effective mass customization. For example, earlier research (Swaminathan et al. 1998) describes the use of postponement in technology industries where semi-finished "vanilla boxes" are produced and stored, delaying differentiation in a short lead time process. Importantly, these semi-finished products impart a high degree of commonality into the product and its components, providing the means for efficient production of mass customized products.

But not all supply chains with common components perform well. In earlier research (Meixell and Wu 1998), factors that influence manufacturing supply chain performance in a collaborative planning environment are identified and analyzed. The factors include commonality of components, the relationship between setup and inventory costs, the planned level of design capacity, and the pattern of end item demand. In this chapter, we give particular attention to manufacturing environments that have been structured to meet the needs of mass customized products and in particular to those that deploy component commonality.

In this research, our objective is to illustrate the impact of collaboration on the production and inventory costs in a supply chain with high component commonality. We associate postponement of product differentiation with component commonality, and ask how commonality influences the bullwhip effect in a collaborative production planning environment. In particular, we investigate which factors have a significant influence on supply chain performance, and what are the interactions between these factors. The insights that result from this investigation suggest a supply chain strategy that could lead to better supply chain performance.

This chapter is organized into five sections. The next section summarizes concepts, models, and results from earlier research on collaboration in the supply chain and on the impact of commonality on production performance. Section 3 describes and then presents the results of a study that addresses these questions, followed by Section 4 that addresses the role of information technology in collaborative manufacturing. Finally, Section 5 is a discussion of managerial impacts of this type of collaboration. We close with a discussion of the research issues and future directions for this line of work.

2. COLLABORATION IN THE SUPPLY CHAIN

In its most general sense, collaboration occurs when two or more companies work together in a mutually supportive and beneficially inclusive fashion to achieve a common objective. Each partner subscribes to the goal of improving performance at the supply chain level, and all partners benefit by participating.

Collaboration is commonly achieved by one of two means – the first is the exchange of data and information (Keskinocak and Tayur 2001). Wal-Mart, for example, exchanges data and information with its suppliers through Retail Link. But sharing selected data with trading partners is a limited form of collaboration. An important second step is to concurrently develop plans for both products and production. We generally think of supply chains as consisting of product, information and monetary flows – this second step focuses on joint planning to integrate these flows not only within a firm, but also across organizational boundaries. Businesses in a collaborative supply chain take actions that increase the total supply chain profit, and avoid actions that improve local profits but diminish overall performance (Chopra and Meindl 2001). The goal is to reduce inefficiencies but at the same time improve responsiveness, by treating multiple organizations in the entire supply chain as a system. In some sources collaboration is further defined as multi-organization optimization over a well-defined and mathematically described process (Fleischmann et al. 2000, Simchi-Levi et al. 2003)

Collaboration in the supply chain can drive benefits in flexibility, lead time reduction, and cost reduction. In a mass customization environment, these benefits are essential to achieving the dual objectives of low price and delivery promptness. The Swedish retailer IKEA collaborates closely with its suppliers to reduce supplier costs and ultimately product price (Margonelli 2002). Sport Obermeyer achieves a great deal of flexibility and reduced lead times by collaborating closely with both customers and suppliers (Fisher et al. 1994). Dell Computers reduces the production lead time that would be incurred in their build to order operation also through close collaboration with customers and suppliers (McWilliams 1997).

2.1   Collaborative processes in the supply chain

Numerous types of collaborative systems have emerged in practice. A useful way to organize them is by the processes around which the collaboration is organized and which it supports. Supply chains may collaborate on the processes that bring a new or revised product to market or on the processes associated with fulfilling orders through a production process – such as product design, inventory management, transportation planning and production and material planning. These processes support a mass customized product and may be essential in achieving the dual objectives of price and delivery promptness, but in practice are not limited to this environment.

Perhaps the best-known type of supplier collaboration is in product design. When collaborating on product design, a manufacturer involves the supplier in the design process to incorporate both technology ideas and information relative to production costs and constraints for alternative product designs. Firms that integrate suppliers into the design process benefit from reduced material costs, increased quality, and reduced product development time (Monczka et al. 1997). Collaboration between a manufacturer and suppliers is perhaps even more essential in a supply chain with mass customized product. Although the topic of including the supplier in the product design decision isn't raised, a number of sources (Fine 2000, Garg 1999) address the importance of designing product and the associated production processes to support the overall performance of the supply chain.

Supply chains may also collaborate on the demand planning process, which contributes to price and delivery time objectives associated with mass customized products. For example, collaborating on promotion planning, forecasting and replenishment (CPFR) involves an exchange of information, generally between a retailer and their suppliers. In a traditional relationship, the retailer may simply inform the supplier about promotion plans, but in CPFR the retailer and supplier integrate their information on promotion timing, finished goods inventory, production plans, production capacities, and raw material availability to generate a superior forecast and plan that supports promotions and serves as a basis for replenishments to the retailer (McKaige 2001, Sherman 1998).

Supply chains may also collaborate on transportation planning, which contributes to the price and delivery promptness objectives associated with mass customized products. Transportation collaboration has been shown to reduce transportation time and variability in a pilot study that involved Wal-Mart, Procter and Gamble, and one of their carriers, J.B. Hunt (Russell 2002). Earlier work in the automotive industry on backhaul planning in the transportation of finished goods also illustrates the impact of collaborating transportation between multiple shippers and carriers (Kunkel 1984).

Supply chains may also collaborate on replenishment planning, which contributes to the price and delivery promptness objectives associated with mass customized products. These types of systems are most common in the retail industry, and were some of the earliest forms of collaboration in the supply chain, and are called Vendor Managed Inventory (VMI), Vendor Managed Replenishment, or Continuous Replenishment. The retailer provides sales history and ongoing POS data to the supplier of the items, who in return decides when and how much to ship to each of the retailer locations. VMI improves flexibility for the supplier who can choose to produce product consistent with the demands of other customers as well as production capacities and costs, and so reduces costs and often increases revenues for the supply chain (Davis 1995).

Collaboration in production and material planning also influences performance in a supply chain and so contributes to the price and delivery promptness objectives associated with mass customized products. Firms may integrate production planning by implementing Advanced Planning Systems (APS) which allow supply constraints to be folded into the customer's production schedule through the Available-to-Promise (ATP) construct and alerts suppliers when problems or infeasibilities occur (Kilger and Schneeweiss 2000, Rohde 2000). Other authors investigate this environment and find that not collaborating with suppliers introduces significant cost penalties, especially when the suppliers are inflexible and impose long lead times on the supply chain (Wei and Krajewski 2000). Inman and Gonsalvez

(1997) also study the impact of failure to synchronize production schedules along an automotive supply chain, and argue that integration of this type is necessary to achieve high performing supply chains. Reviews of the literature (Beamon 1998, Erenguc et al. 1999, Ganeshan et al. 1999) indicate that much of the work in production and material collaboration address inventory coordination. Lin et al. (2000) also address inventory coordination, and argue the need for collaboration in an extended enterprise in a technology industry, and present a model to coordinate safety stocks throughout the supply chain so as to minimize total inventory. Graves et al. (1988) also present a model for supply chain inventory optimization and report on an application that supports the need for integrating material stocks across multiple tiers in a supply chain. It is this process with collaboration in production and material planning that we focus on in this research.

2.2   Component commonality

At its most fundamental level, product variety influences individual production facilities in a supply chain in terms of cost, responsiveness and quality. As product variety increases, penalties are incurred in manufacturing plants in batch related activities and in product-sustaining activities (Cooper and Kaplan 1991). With higher variety, lower volume over more products translates into more frequent setups and so more setup time that detracts from both equipment and labor productivity. Product-sustaining costs include the engineering change process, as product changes are more numerous with high variety. Other costs that increase with high variety include materials related costs for ordering, expediting, shipping, receiving, and storage; and quality-related costs for managing conformance to specifications. Fisher et al. (1994) studied the impact of product variety in the automotive industry and found that increased variety increases overhead costs because more effort is required for demand forecasting; more inventory is required because of the higher degree of forecast error due to a disaggregation of demand, more complex scheduling, and more frequent engineering changes.

The link between production facilities in the supply chain relative to product variety is the product structure, which defines the degree to which components are shared between products purchased by customers in the supply chain. So, the degree to which components are shared is a determinant in the influence of product variety on supply chain performance. Randall and Ulrich (2001) show that higher variety drives higher production costs in the bicycle industry supply chain without component commonality because it is more difficult to achieve minimum efficient scale production when components are more numerous. Fisher and Ittner (1999) investigate component sharing as a tactic for achieving scale economies in production while maintaining a profitable level of product variety. Benton and Krajewski (1990) address component commonality in their study on supplier performance and conclude that lead time uncertainty can be reduced with higher levels of commonality, that intermediate inventories may increase, and that quality of materials may decrease.

3. THE INFLUENCE OF COMPONENT COMMONALITY ON SUPPLY CHAIN COSTS

In this section, we report the results of a computational study that addresses how component commonality influences supply chain performance. We use demand amplification – also known as the bullwhip effect - as a measure of supply chain performance. Specifically, we expand upon an earlier multi-factor study on supply chain performance in a collaborative production planning environment (Meixell and Wu 1998). The earlier study found that setup costs, capacity utilization, component commonality, and in some cases, end item order pattern all impact demand amplification across the supply tiers. Component commonality is found to improve supply chain performance relative to supply chains without component commonality. In fact, there are cases where demand variability is reduced tier to tier, suggesting an inverted bullwhip effect or de-amplification of demand variation in a supply chain.

The individual simulation results are extracted from the comprehensive study and are examined here as a secondary experiment in the context of component commonality. Accordingly, we investigate demand amplification between the first two tiers depicted in Figure 7-1 only whereas the original study included amplification effects across a five tier supply chain. We then relate these computational results to analytical findings from (Wu and Meixell 1998).

3.1 Modeling Collaborative Production Planning

A production planning model is developed and implemented here to study collaborative manufacturing supply chains where cost, capacity, and demand information are shared between supply chain partners. This information is then used to integrate production planning decisions across multiple enterprises and facilities. In this type of production planning system, the central decision is to determine the production quantity of each item in each time period of the planning horizon. The model is a multi-level, multi-period lot-sizing model that generates a solution that is descriptive of demand propagation in collaborative manufacturing supply chains. The production planning activity in the supply chain is dependent on the production plan at previous tiers, on the order processing decisions made at each successive tier, and on the cost and constraint information made available by key suppliers throughout the supply chain. Importantly, we model these decisions as an integrated system that suggests that the tiers do indeed collaborate by sharing information and acting to support the performance of the overall supply chain. This model and the heuristic used to solve it are further described in (Meixell 1998).

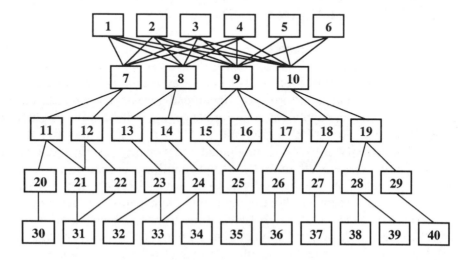

Figure 7-1. Component Commonality in the Supply Chain

Demand amplification is used in this study to measure supply chain performance in an aggregate fashion. How demand propagates through tiers in a supply chain defines a fundamental characteristic and a primary performance driver in supply chains. Demand amplification as the bullwhip effect has been used in earlier work as a surrogate for supply chain performace (Lee et al. 1997) and profitability (Metters 1997). The bullwhip effect is known to increase manufacturing, inventory, and transportation costs, as well as have a detrimental effect on product availability in build-to-stock systems and delivery reliability in build-to-order systems (Chopra and Meindl 2001).

In particular, more inventory is held, production overtime is incurred, transportation is more often expedited, and due dates are more frequently missed. In this study, demand amplification is measured as the growth in the coefficient of variation from the first to the second tier in the supply chain.

3.2   Experimental Design

In this experiment we used a multi-factor design with treatment levels as defined in Table 7-1 to examine the magnitude and direction of the factor effects and their interactions. These varying combinations of levels of the context variables are tested by populating the experimental design with problem instances that are solved using the collaborative planning model described earlier. We focus on three factors – setup costs, targeted capacity utilization, and end item demand patterns. The levels for the setup costs factor are set as "low" and "high" relative to inventory costs. Likewise, targeted capacity utilization is set at a relaxed level (50% of peak demand) and a tight level (90% of peak demand). Three demand patterns are tested – a random pattern, a level pattern, and a balanced or negatively correlated pattern as illustrated in Figure 7-2.

Table 7-1. Simulation Factors

	Level 1	Level 2	Level 3
Setup Cost Profile	Low	High	—
Targeted Capacity Utilization	50%	90%	—
End-Item Demand Pattern	Random	Level	Negatively Correlated

All combinations of these factor levels are simulated as instances that represent scenarios, and we examine both the main and the interaction effects of the selected factors on the supply chain performance using the demand amplification as a measure. Specifically, we evaluate how each of the identified factors influences demand amplification, defined as the growth in the coefficient of variation (CV) of demand as it translates through the supply chain. The response variable then is the difference in the CV at the second and first tiers in the chain.

A well-constructed test database developed by Tempelmeier and Derstroff (1996) was used to populate the experimental design. This dataset contains 1200 randomly generated problem instances that vary systematically in product and operation structure, setup time, time between order profile, design capacity, and setup cost. For each required treatment combination, the instances needed for this study were selected at random to fit the experimental design. The problems each consist of 40 items with 16 periods of demand, produced at 6 different facilities, distributed across the tiers in the supply chain. In all cases, lead-time is set equal 1 and the initial inventory is 0.

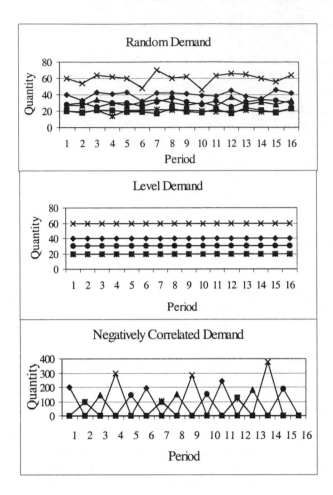

Figure 7-2. End-Item Demand Patterns

3.3   Computational Results

The simulation produced a set of results from which we derive the coefficient of variation (CV) of demand for each level in the five tier supply chains analyzed in this study. Figure 7-3 maps several of the problem instances, showing how the CV of demand changes. The charts show that the bullwhip effect as measured by demand amplification may still occur in supply chains with component commonality. Under select conditions, however, the bullwhip effect is absent, as the CV of demand remains constant or actually diminishes with increasing depth into the supply chain. This de-amplification provides favorable conditions for the supply chain as it helps to reduce production and distribution related costs, aiding in the achievement of the dual objectives of fast delivery and reasonable price.

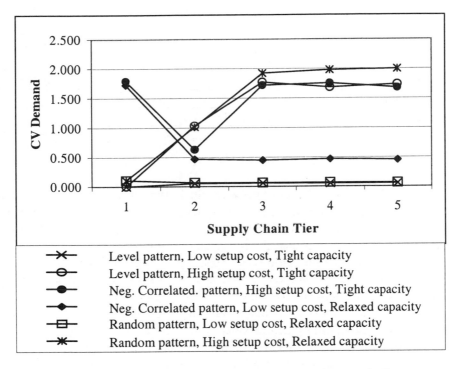

Figure 7- 3. Demand Amplification in Selected Manufacturing Supply Chains

Table 7-2 presents the analysis of variance (ANOVA) results and summarizes the main effects and interactions for demand amplification. The tables show that two of the three factors tested, order pattern and setup costs are significant in excess of the 95% level of significance, and the third factor, design capacity, is significant at a lower level of significance.

Table 7-2. Results from the Analysis of Variance

Source	Sum of Squares	Df	Mean Square	F-Ratio	P-Value
Main Effects					
A: Order Pattern	7.337	2	3.669	435.4	0.002
B: Setup Cost	1.428	1	1.428	169.5	0.006
C: Capacity	0.094	1	0.094	11.1	0.079
Interactions					
AB	0.500	2	0.250	29.7	0.033
AC	0.177	2	0.089	10.5	0.087
BC	0.001	1	0.001	0.14	0.742
Residual	0.017	2	0.008		
Total	9.554	11			

Table 7-3 further illustrates these results in the form of a means table that shows the direction and magnitude of demand amplification for each of the factors. Through inspection of the means, we draw a few interesting conclusions. First, we see that the presence of component commonality alone will not eliminate the bullwhip effect as a number of the level combinations exhibit a positive amplification. However, a few of the factors at select levels either hold CV of demand constant or reduce it. One such factor is the order pattern, which influences demand amplification in supply chains with component commonality when production plans are coordinated. From Table 3 it is clear that the most favorable demand pattern is the negatively correlated case, where the demand amplification takes on a negative sign signifying that de-amplification occurs. The means table also indicates that on average, both the random and the level order patterns exhibit increasing amplification although less with the random than with the level order pattern. A level order pattern is not beneficial for the supplier then, even though it may be beneficial to the first tier manufacturer.

A second observation from this analysis pertains to the setup cost factor, where we see that low setup cost tends to drive less demand amplification

Table 7-3. Experimental Results – Main Effects

Level	Count	Mean
Grand Mean	12	-0.187
Order Pattern		
Level	4	0.538
Negatively Correlated	4	-1.273
Random	4	0.175
Setup Cost		
Low	6	-0.532
High	6	0.158
Targeted Capacity Utilization		
Tight Capacity	6	-0.275
Relaxed Capacity	6	-0.098

than high setup cost. In fact, negative demand amplification results for some of the low setup cost cases. In practice, a high setup cost motivates planners to batch customer orders and so the CV of demand increases as orders are modified by the order management process. When setup costs in the supply chain are low relative to inventory costs however, one of two results occurs. In some cases, demand variation is constant. In others, demand variation is reduced and a de-amplification effect results.

A third result pertains to the design capacity factor. Here, the favorable condition is high or tight design capacity utilization. There is little opportunity in this case for the manufacturer to batch multiple orders to achieve regardless of setup costs, which results in a smoothing effect and a reduced level of amplification in the supply chain, as noted in our earlier paper (Meixell and Wu 1998).

Note that the ANOVA results in Table 7-2 suggest a strong interaction between order pattern and setup cost. In Figure 7-4, we investigate this interaction more closely. This graphic shows that for both the random and the level end item demand patterns, demand tends to be amplified from the first to second tier, but the setup cost influences the degree to which this phenomenon occurs. As illustrated earlier and observed here again, high setup cost tends to increase demand amplification in the supply chain. Interestingly, for the negatively correlated case, this difference is nullified so that regardless of the setup cost, there is strong and negative amplification when the end item order patterns are balanced so that a negative correlation exists.

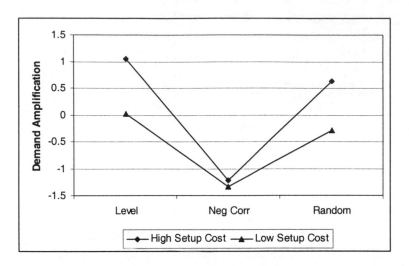

Figure 7-4. Interaction chart between order pattern and setup costs

These experimental results on component commonality and demand correlation are supported analytically in (Wu and Meixell 1998). The research reported in this earlier paper finds that when n end-items with demands correlated as ρ use the same component, demand is amplified by a factor of $\left(\sqrt{\dfrac{1+n\rho-\rho}{n}}-1\right)$ when orders emanating from the first tier are batched for production in the second tier. Note that the only cases of non-negative amplification, $\left(\sqrt{\dfrac{1+n\rho-\rho}{n}}-1\right)\geq 0$, are when $\rho=1$, $n=1$, or both. The amplification factor is otherwise negative, consistent with the de-amplification effect observed in the experiment with negatively correlated demands. In these cases, the CV at the component level is reduced relative to the CV at the end-item level.

4. DISCUSSION

These results have important implications for achieving the dual objectives of price and speed in a supply chain. We see that using common components, a common practice in mass customization, does not alone mitigate the bullwhip effect in a manufacturing supply chain. Collaborating in the fashion described above – i.e. by coordinating production plans such that costs and constraints from across the supply chain are taken into account in a collaborative fashion - provides an opportunity for performance improvement in manufacturing supply chains. This result occurs in a specific context, however. When setup costs are low relative to inventory costs, there is little incentive for amplification and in fact we find that de-amplification occurs in the supply chain between the top two tiers. Additionally, de-amplification occurs when the end item demand patterns are negatively correlated – regardless of setup cost.

This de-amplification effect, then, is most prevalent in the case where the end item demands are well balanced as exhibited by a negative correlation. This condition may occur naturally in the market, but is more likely to occur when induced. These patterns could be induced by a variety of mechanisms, such as a demand management mechanism that prices or provides other motivation to customers to place orders according to this pattern. Order pattern can also be influenced via scheduling mechanisms with practices such as order leveling and order balancing, both common in industry. If a particularly large number of orders with a critical component are received for production in a particular time period, some of these orders can be delayed and scheduled in the next time period (incurring late or expedited delivery), or in the previous time period if known with sufficient lead time (incurring inventory costs). On the other hand, order balancing can be accomplished by scheduling a specific version for a critical component in every other, or possibly every third period, depending on its relative frequency. Again, inventory or late delivery charges may apply.

The de-amplification effect also addresses the incentive and gain sharing obstacle to collaboration in a supply chain. Suppliers often have little incentive to participate in collaboration schemes when they perceive that there is little gain for their own organization (Forger 2002, Kumar 1996, Lee and Whang 1999). In an optimized network, there are typically both losers and gainers that naturally emerge when performance gains are not re-distributed. Some trading partners will experience decreased costs while others incur increases relative to the un-coordinated network, even when the network as a whole enjoys a performance improvement. When de-amplification occurs in a network, however, both tiers benefit. The top manufacturing tier benefits from the balanced demand pattern that requires less inventory, excess capacity, and overtime. And the supplier benefits by having a demand pattern that carries even less variation than end item demand pattern.

5. INFORMATION TECHNOLOGY IN THE COLLABORATIVE MANUFACTURING SUPPLY CHAIN

Plans are developed by sales, manufacturing, materials management and engineering functions within an enterprise, so for inter-enterprise collaboration to be successful, a single internally consistent plan is important.

A second requirement addresses data transfer between organizations. Historically, trading partners have transmitted transaction files to each other for decades using EDI on private networks (Segev et al. 1997). These transactions – ANSI standardized – include material releases, transportation instructions, order and shipping tracking information, and shipment acknowledgements. These transactions typically occurred between the manufacturers in the top tiers of the supply chain, and are tremendously valuable in reducing transaction costs associated with managing large manufacturing networks. Traditional EDI also has limitations - EDI on private networks is costly and sometimes prohibitively so for smaller companies. This is problematic because it excludes potentially valuable trading partners from the network. For example, a small supplier who can handle variability in demand inexpensively brings value to the network that a large, EDI-enabled supplier may not be able to.

The information required for collaboration between trading partners, however, requires more than simple transaction file transmission. One view of e-commerce communication in the supply chain presents a four level categorization (Simchi-Levi et al. 2003). These levels start with simple one-way communication either from a buyer to a supplier or vice versa, and can be accomplished via email, or web browsing. The second level provides data base access from one trading partner and for another and may require personal or tailored information. For example, procurement exchanges such as e-bay are in this category. A third level is data exchange between buyer and supplier and might be accomplished using EDI or XML programs. The fourth level communications enable shared business process decision making, such as the collaborative manufacturing processes. At this level, IT standards are essential to support a many to many communication structure. Web services – applications of nearly any type made available to remote machines over the Internet - also support this fourth level of communication (Fletcher and Waterhouse 2002). Collaboration between enterprises is not practical without these technologies.

Collectively, these collaborative systems are called advanced planning systems (APS), and are built to support the integration of decisions both within and between enterprises in a supply chain. These systems are a set of inter-related but individual modules that coordinate tasks in procurement, production, distribution, and sales. In each of these functional areas, the modules offered by software developers address one set of particular problems. These software companies include i2, SAP (APO), Manugistics, PeopleSoft, J.D. Edwards, Oracle and others. In production, for example, the problems of plant location, production system design, master production scheduling, capacity planning, lot sizing, machine scheduling, and shop floor control each have a module with appropriate decision technology to support the decision process (Fleischmann et al. 2000). Cisco has integrated its inter-enterprise business processes in this way through Internet-enabled information technology (Kraemer and Dedrick 2002).

6. CONCLUSIONS

This research investigates an important coordination problem for supply chains in a manufacturing environment engineered to achieve mass customization and provides valuable managerial insights concerning collaboration in such systems. We simulate multi-tier supply chain production coordination using a lot-sizing based production planning model, and study how demand propagates in this complex manufacturing environment. As such, the findings of this work links the effects of order batching, design capacity, end item order patterns with supply chain performance.

As a result, we see a number of managerial implications for supply chains where postponement strategies have been utilized such that components are common. First is that end item order pattern, relative setup cost, and design capacity, roughly in that order, have a significant influence on supply chain performance in terms of demand amplification. Decisions concerning these investments should incorporate the benefits associated with improved demand propagation across the supply chain. Perhaps additional investment in relative setup should be made to reduce the per setup cost, reducing demand amplification in the supply chain.

This study also illustrates the managerial implication of end item order patterns in achieving improved performance in the top tiers of manufacturing supply chains. No doubt, there is a natural tendency in some markets for positive correlation to occur in an un-altered order management process. These patterns can be modified through demand management techniques structured to induce negative correlation in end item order patterns with a variety of pricing and scheduling mechanisms described earlier. It is also important to recognize that although level demand patterns may be beneficial to the top tier manufacturer, it is not beneficial to the supply base in a general setting as demonstrated in the analysis here.

Additional research on this topic is needed to further investigate the impact of product variety and component commonality on the collaborative manufacturing supply chain. This chapter shows the influence of component commonality on the aggregate demand amplification measure, but future research should expand on these results and investigate the specific costs and responsiveness metrics that are influenced in the supplier operation as component commonality increases. Managers benefit from knowing the degree to which these benefits might be achieved in part because they can then be compared to the costs of building collaborative systems. Also, it would be useful to investigate how other collaborative processes – such as forecasting and transportation planning – influence the bullwhip effect in general and the specific supplier performance metrics in particular.

REFERENCES

1. Beamon, B.M. Supply chain design and analysis: models and methods International. Journal of Production Economics 1998; 55: 281-294.
2. Benton, W.C., and Krajewski, L. Vendor performance and alternative manufacturing environments. Decision Sciences. 1990; 21: 403-415.
3. Chopra, S., and Meindl, P. Supply Chain Management: Strategy, Planning and Operations. Upper Saddle River, New Jersey: Prentice Hall, 2001.
4. Cooper, R., and Kaplan, R.S. The Design of Cost Management Systems: Englewood Cliffs, NJ: Prentice Hall, 1991.
5. Davis, D. State of a New Art. Manufacturing Systems 1995; 13: 25-30.
6. Erenguc, S.S., Simpson, N.C., and Vakharia, A.J. Integrated production/distribution planning in supply chains: An invited review. European Journal of Operational Research 1999; 115: 219-236.
7. Fine, C.H. Clockspeed-based strategies for supply chain design. Production and Operations Management 2000; 9/3: 213-221.
8. Fisher, M., and Ittner, C. The impact of product variety on automobile assembly operations: Empirical evidence and simulation analysis. Management Science 1999; 45/6: 771-786.
9. Fisher, M.L., Hammond, J., Obermeyer, W., Raman, A. Marking Supply Meet Demand in an Uncertain World. Harvard Business Review) 1994; May-June: 83-93.
10. Fleischmann, B., H. Meyr, and M. Wagner, "Advanced Planning" in: Supply Chain Management and Advanced Planning: Concepts, Models, Software and Case Studies, H. Stadtler and C. Kilger (eds.). New York: Springer-Verlag, 2000.
11. Fletcher, P., Waterhouse, M. Web Services Business Strategies and Architectures. Birmingham, U: Expert Press, 2002.

12. Forger, G.R. The problem with collaboration. Supply Chain Management Review 2002; 6/2: S56-57.
13. Ganeshan, R.E., Magazine, M.J., and Stephens, P. "A Taxonomic Review of Supply Chain Management Research," in: Quantitative Models for Supply Chain Management, S. Tayur, R.E. Ganeshan and M.J. Magazine (eds.). Boston: Kluwer Academic Publishers, 1999.
14. Garg, A. An application of designing products and processes for supply chain management. IIE Transactions 1999; 31/5: 417-439.
15. Graves, S., Kletter, D., and Hetzel, W. A dynamic model for requirements planning with application to supply chain optimization. Operations Research 1988: 46/S3: S35-S49.
16. Inman, R.R., and Gonsalvez, D.J.A. The causes of schedule instability in an automotive supply chain. Production and Inventory Management Journal 1997: 38/2: 26-31.
17. Keskinocak, P., and Tayur, S. Quantitative Analysis for Internet-Enabled Supply Chains. Interfaces 2001: 31/2: 70-89.
18. Kilger, C., and Schneeweiss, L. "Demand Fulfillment and ATP" in: Supply Chain Management and Advanced Planning: Concepts, Models, Software and Case Studies, H. Stadtler and C. Kilger (eds.). New York: Springer-Verlag, 2000.
19. Kraemer, K.L., and Dedrick, J. Strategy use of the Internet and e-commerce: Cisco Systems. The Journal of Strategic Information Systems 2002: 11/1: 5-29.
20. Kumar, N. The Power of Trust in Manufacturer-Retailer Relationships. Harvard Business Review 1996; November-December.
21. Kunkel, M. J. Approximating Scale Economies in General Motors Vehicle Shipping: Impacts on Truck Backhauling Opportunities. Research Report TN-306, 1984, General Motors Research Laboratories, Warren, Michigan.
22. Lee, H.L., Padmanabhan, V., and Whang, S. Information Distortion in a Supply Chain: The Bullwhip Effect. Management Science 1997; 43: 546-558.
23. Lee, H. L. Whang, S. Decentralized Multi-Echelon Supply Chains: Incentives and Information. Management Science 1999; 45/5: 633-642.
24. Lin, G., Ettl, M., Buckley, S., Bagchi, S., Yao, D.D., Naccarato, B.L., Allan, R., Kim, K., and Koenig, L. Extended-Enterprise Supply-Chain Management at IBM Personal Systems Group and Other Divisions. Interfaces 2000; 30/1: 7-25.
25. Margonelli, L. How IKEA designs its sexy price tags. Business 2.0 2002; 3/10: 106.
26. McKaige, W. Collaborating on the Supply Chain. IIE Solutions 2001; March: 34-37.

27. McWilliams, G. Whirlwind on the Web. Business Week 1997; April 7: 132-136.
28. Meixell, M. J. Modeling Demand Behavior in Manufacturing Supply Chains. Department of Industrial and Systems Engineering, Lehigh University, Bethlehem, Pennsylvania, 1998: un-published Doctoral dissertation.
29. Meixell, M. J., Wu, S. D. Demand Behavior in Manufacturing Supply Chains: A Computational Study. IMSE Technical Report 98T-008, Lehigh University, Bethlehem, Pennsylvania.
30. Metters, R. D. Quantifying the bullwhip effect in supply chains. Journal of Operations Management 1997; 15/2: 89-100.
31. Monczka, R., Ragatz, G., Handfield, R., Trent, R., and Frayer, D. Supplier Integration into New Product Development: A Strategy for Competitive Advantage - The Global Procurement and Supply Chain Benchmarking Initiative. The Eil Broad Graduate School of Management, Michigan State University, East Lansing, Michigan, 1997.
32. Randall, T., Ulrich, K. Product Variety, Supply Chain Structure, and Firm Performance: Analysis of the U.S. Bicycle Industry. Management Science 2001; 47/12: 1588-1604.
33. Rohde, J. "Coordination and Integration," in: Supply Chain Management and Advanced Planning: Concepts, Models, Software and Case Studies, H. Stadtler and C. Kilger (eds.). New York: Springer-Verlag, 2000.
34. Russell, D. M. Integrating CTM & CPFR: A Proposed Process and Tactics for Managing the Broader Supply Chain Collaboration," in: Collaborative Planning, Forecasting, and Replenishment, The Harvard Business School Press, Cambridge, Massachusetts, 2002.
35. Segev, A., Porra, J., and Roldan, M. Internet-based EDI strategy. Decision Support Systems 1997; 21: 157-170.
36. Sherman, R.J. Collaborative Planning Forecasting & Replenishment (CPFR): Realizing the Promise of Efficient Consumer Response Through Collaborative Technology. Journal of Marketing Theory and Practice 1998; 6/4: 6-9.
37. Simchi-Levi, D., Kaminsky, P., and Simchi-Levi, E. Designing and Managing the Supply Chain: Concepts, Strategies and Case Studies. New York: (2nd ed.) McGraw-Hill/Irwin, 2003.
38. Sodhi, M. S. Applications and Opportunities for Operations Research in Internet-Enabled Supply Chains and Electronic Marketplaces. Interfaces 2001; 31/2: 56-69.
39. Swaminathan, J. M. Tayur, S. R. Designing Task Assembly and Using Vanilla Boxes to Delay Product Differentiation: An Approach for Managing Product Variety" in: Product Variety Management: Research

Advances, T.-H. Ho and C.S. Tang (eds.). Massachusetts: Kluwer Academic Publishers, Norwell, 1998.

40. Tempelmeier, H., Derstroff, M. A Lagrangean-based heuristic for dynamic multilevel multiitem constrained lotsizing with setup times, Management Science 1996. 42:5, 738-757.

41. Wei, J., Krajewski, L. A model for comparing supply chain schedule integration approaches. International Journal of Production Research 2000; 38/9: 2099-2123.

42. Wu, S. D., Meixell, M. J. Relating Demand Behavior and Production Policies in Manufacturing Supply Chains. IMSE Technical Report 98T-006, Lehigh University, Bethlehem, Pennsylvania.

CHAPTER 8

SIMULATION MODELING USING AGENTS FOR MASS CUSTOMIZATION

Manfredi Bruccoleri

University of Palermo

Abstract: Reconfigurable enterprises and production networks - virtual structures of manufacturing enterprises that coalesce and vanish in response to a dynamic marketplace – are one of the main enabler paradigm for mass customizable production. The design and coordination of production networks to changing market dynamics are challenging tasks. Simulation provides the ability to develop prototypes of such complex systems cost effectively and in shorter duration. This Chapter describes how simulation can be utilized in conjunction with multi agent system techniques to develop realistic prototypes of complex decision support systems for the design of coordination strategies in large-scale mass customization production systems.

Keywords: Multi agent systems, production networks, simulation.

1. INTRODUCTION

This Chapter focuses on the topics related to the use of multi agent systems and simulation in the design, planning and control of manufacturing systems in the new era of mass customization.

The addressed subject is indeed very interesting, due to the contemporary nature of the analyzed problem, both at research and at industrial levels. Certainly, the current market requirements have pushed the enterprise management efforts towards the challenge of identification of a production and organization response to the industrial need arising from the new paradigm of mass customization. Mass customization requires synthesis between mass production and the production of highly customized products (Pine et al. 1993). Small and medium manufacturers are forced to follow a strategy of differentiation, as the production of small quantities is predominant. The level of output is clearly insufficient for cost leadership and enterprise's need to focus on a more efficient production process.

Scientific papers and workshops, as well as Government and Industry expectations seem to acknowledge that the challenge keyword is "reconfiguration" (US National Research Council 1998). Enterprise reconfiguration, product reconfiguration, production system reconfiguration, control software reconfiguration are some of the many aspects recognized as critical to remain competitive in the "global" market and in the era of mass customization, where the main requirement is to produce a high variety of customized product at a reasonable cost.

Specifically, two major industrial responses to mass customization can be acknowledged:

- **Reconfigurable manufacturing**. From a strictly manufacturing perspective, the most agreed upon technological response to mass customization is connected with the concepts of modularity and reconfigurability of production systems, machine tools, material handling systems, a7nd control software modules. Reconfigurability is an engineering technology that deals with cost-effective and quick reaction to market changes (Koren et al. 1999) and modularity is the key to achieve low cost customization (Pine et al. 1993). Goldhar and Jelinek 1983, discussed modularity in production as a means to partition production to allow economies of scale and scope across product lines. Reconfigurable manufacturing systems (RMS), whose components are reconfigurable machines and reconfigurable controllers, as well as methodologies for their systematic design and rapid ramp-up, are the cornerstones of this new manufacturing paradigm. Indeed, even though they don't offer the high level of flexibility that is proper for a FMS, they allow quick adjustment of production capacity and functionality conducive to produce high mix of products, without incurring high costs of the FMS that are designed to manufacture "every volume" and "every mix". In other words, in order to satisfy the new requirements of mass customization, i.e., the production of customized products in a cost-effective manner by achieving volume-related economies, the key is not to adopt general purpose manufacturing resources, but systems designed for a specific manufacturing objective that are easy to modify and reconfigure in order to respond to a future change in the manufacturing objective.

- **Reconfigurable enterprise**. From an organizational point of view, production network (PN) is considered the industrial response to challenges posed by this new era, characterized by the global market and impressive advances of information and communication technology (Wiendhal and Lutz 2002). Market globalization, indeed, has offered to firms the possibility to split geographically their production capacity. Business opportunities lead firms to work together in temporary organizations. In the same firm, business units behave as autonomous profit centers and compete with each other for production capacity allocation. One of the visions for manufacturing in 2020 is that the structure and identity of firms will radically change to encompass virtual structures that will coalesce and vanish in response to a dynamic marketplace (US National Research Council 1998). Organizations are not able to change their core competencies fast enough to take advantage of meaningful opportunities, so they will have to form alliances. In this way, mass customizable production affects not only the structure of the enterprise but also of its supply chain because it must possess organizational features that allow an easy partnership of members into production networks. The PN members need to be properly coordinated in order to achieve reduction in lead times and costs, alignment of interdependent decision-making processes, and improvement in the overall performance of each member, as well as the PN.

In this context, the design, operation management, and the coordination of reconfigurable manufacturing systems and PNs are challenging tasks. However, it is essential to perform risk-benefits analysis of the design, management, and coordination alternatives before making final decisions. Simulation as a virtual manufacturing tool provides the ability to develop prototypes of such complex networks and systems cost effectively and in shorter duration (Swaminathan et al. 1998). Since early 1990s, simulation began to mature, when many small firms embraced this tool. Better animation, ease of use, fast computers, easy integration with other packages, and the emergence of commercial simulators have all helped simulation become a standard tool in many companies, as well as a widespread tool used in production research (Virtual Manufacturing Technical Workshop 2001).

As far as the design of coordination activities in PN and operation management strategies in RMS are concerned, multi agent system (MAS) techniques have been largely used for their suitability in modeling complex systems involving multiple autonomous agents with internal knowledge and reasoning engines which communicate and negotiate with each other by exchanging messages according to specific negotiation protocols (Smith 1980). The aim of this Chapter is to describe how simulation can be utilized in conjunction with MAS technology for modeling, analyzing, testing and verifying coordination and management strategies in PN prototypes. Concerning issues and discussions of using simulation and MAS techniques for the design of operation management strategies and coordination of reconfigurable manufacturing system, the author suggests that the reader refer to Bruccoleri and Pasek 2002, and Bruccoleri et al. 2003.

The rest of this Chapter is structured as follows. Section 2, after a brief introduction on production networks, gives an overview on major issues related to the use of MAS technology and simulation in modeling and analyzing PN. Section 3, presents the MAS framework and the simulation model that have been designed for a specific coordination issue in production networks, i.e. the decision making process for allocating customer orders to the members of the PN itself. Finally, a numerical case study is presented in Section 4.

2.   PRODUCTION NETWORKS

A quick overview of published literature on problems and issues facing the new industrial landscape, points towards to strong evidence of the expansion of production networks, distributed enterprises, and virtual enterprises. Although a commonly accepted formal definition for the idea of manufacturing into networks does not yet exist, its impact on the manufacturing industry and research is quite impressive.

In general it can be said that the term production network refers to a temporary or stable coalition among companies in order to achieve a certain goal. Many definitions can be given depending on the type of coalition, on the meaning of the terms "temporary or stable", and on the aim of the coalition itself.

Regarding the coalition (which means the type of partnership that exists among the members of the network) and the duration of the coalition itself, a clear classification can be found in Wiendhal and Lutz (2002). They distinguish supply networks, virtual enterprises, production networks, and clusters according to the level of hierarchy and dependency among the members and the duration of the production network.

More interesting is the distinction that can be given in terms of the objective and aim of the production network. Essentially, three main types of production networks can be considered.

The first type refers to firms that become partners, in order to reach a critical size to be in accordance with market constraints. Mass customization, indeed, if on the one hand could require production of a single product as an extreme result of constant market differentiation, on the other hand, it could call for a very high volume production. This happens, for example, for major components of a modular product whose real differentiation and customization affect only the last phases of the manufacturing process, or simply the assembly phase. In other words, as mass customizable production is frequently achieved by means of product modularization, often the ability to produce a high variety of products relies on different combination and permutation that can be given to the product modular components. The common components, of course, have to be produced in high volume and this may require the collaboration of some partner with whom to split the production volume.

The second type of production network refers to cooperation and alliance among companies of a supply chain in order to optimize the global supply chain. Mass customization, as a competitive strategy, requires that different production types are employed simultaneously with high requirements for inter company interactions (Turowski 2002). Indeed, once the customer has configured its customized product, the manufacturer checks which component type and how much of each component can be produced by the manufacturer and which of these must be acquired, forwarded to a partner, or split with another manufacturer. The supplier or the partner of this manufacturer, then, becomes client or partner of another manufacturer and so on. This leads to a nesting of procurement activities, and finally to a networked production process that covers multiple companies and plants. In other words, the procurement follows the production of individual products or a partition of the customer order volume by flexible and automated groups.

Finally, the last type of production network is the cooperation driven by the need to develop a pool of competencies and resources in order to maximize flexibility and adaptability to environmental changes (Martinez et al. 2001). Market opportunities can arise and disappear within a short timeframe and the ability of traditional enterprises to adapt to these changes remains limited. Often, the environment changes more rapidly than enterprises can adapt to it. In a mass customization environment, requests of customized products or even of individual products could represent the only business opportunities to stay competitive. Virtual enterprises are designed to behave in an agile conduct towards market opportunities by sharing of competences and capacity of the partners (Katzy and Dissel 2001).

2.1 Multi agent systems in production networks

In a traditional supply chain, as soon as a customer launches an order to a manufacturer, real-time planning can occur and the manufacturer is able, in a short interval, to confirm or to counter offer a new proposal to the customer in terms of delivery date, volume, price, and quality details. Although, if the requested products are not standardized, as in the case of mass customization environment, this process is not possible and the manufacturer must negotiate with other supplier or with other members of the network concerning, whether or not demand could be covered and how the order should be split and allocated to the single partner. Only after this process, an offer can be sent back to the customer. Even in those cases when a manufacturing resource breaks down, the process of negotiation with other manufacturing plants could be initiated in order to perform manufacturing operations that cannot temporarily be performed by the enterprise with the failed resource. For all of these processes the coordination of production activities within and between members of a production network is necessary.

The basic approach to attempt to solve coordination problems is to build up a centralized form of coordination; however, such forms of coordination consume significant amount of time and resources. Moreover, as the level and the scope of coordination increase, the notion of centralized coordination breaks down at a point where the system complexity reaches its limit.

Multi agent environments have been largely used in decentralized platforms for their suitability in modeling complex systems involving multiple autonomous agents which negotiate with each other by exchanging messages according to specific negotiation protocols. MAS is a branch of distributed artificial intelligence and the term agent represents a software system that has the properties of autonomy, social ability, reactivity, pro-activeness, proper knowledge belief, intention and obligation (Ferber 1999). Agent technologies are spreading very fast in several areas and constitute the base of business to business applications.

It is well known that multi agent systems are nowadays widely utilized in the design of distributed shop floor control systems, because they implement a local control approach, which allows simplifying the design of a complex system and improves the adaptability of the system to modifications of the structure (Shen and Norrie 1999). Nevertheless, their employment is very suitable for modeling supply and production networks in different stages of a network operation and for several design issues (Swaminathan et al. 1998). MAS architecture for managing the virtual supply chains and production networks at the tactical and operational levels views those systems as composed of a set of intelligent (software) agents, each responsible for one or more activities in the network operation and each interacting with other agents in planning and executing their responsibilities (Fox et al. 2000). An agent is an autonomous, goal-oriented software process that operates asynchronously, communicating and coordinating with other agents as needed. Brandolese et al. (2000) propose a multi-agent approach for the capacity allocation problem in a supply chain over the traditional centralized algorithms and demonstrate how MAS is very promising technology for coordinating production planning activities in complex manufacturing system. For the complex coordination issues intrinsic to the network operation, a centralized and hierarchical control approach would need the simplification of many hypotheses in order to prevent cases of chaotic behavior. A deterministic and preventive control cannot, indeed, anticipate all the possible real cases. Unpredictable cases are due to the interaction between manufacturing units or virtual enterprise partners that are led by an upper production control level (Martinez et al. 2001).

Firms in a production network are independent and there is no single authority that governs the entire network collaboration. They exchange information such as customer demands, inventory levels, but do not control each other one-sidedly. Also, intelligent coordination is required for planning and scheduling of production and logistics in a dynamic market environment (Ahn et al. 2003). Agent technology has many desirable features, viz., autonomy, intelligence and collaboration for implementing such coordination. In agent-based architecture, for coordination, functional units of individual participant to the network are modeled as independent and collaborating agents, such as order agents, enterprise agents, logistic agents, and scheduling agents. The agent framework is usually used to show how to coordinate various activities of the real actors of network under different conditions.

2.2 Simulation in production networks

Within an agent-based framework, simulation becomes necessary during the phases of design and test of negotiation models that rule the distributed control of networks. In the Italian research project PRIN2001 (Process and production planning in manufacturing enterprise networks), simulation is used to test and validate negotiation model for supplier selection and order bidding in a supply chain-oriented e-marketplace (Cantamessa et al. 2002).

In general, the design, coordination, and management of complex PN, as well as the selection of the business opportunity, the choice of partners, the design of the inter-company information system, and production planning and control are all critical tasks. Simulation can be used for modeling PN and complex supply chains at different abstraction levels and for a variety of purposes. For example, the selection of partners (one of the preliminary phases of production network design) entails the careful analysis of their core competences. The partner's coherence with network's strategy and needs is a crucial activity during such phase. Also important is the amount of effort needed to coordinate and integrate a member within the network. Simulation could be essentially beneficial in choosing partners. As an example, Fuji et al. (2000), by means of simulation, studied the effect of participating in a network through variations in the manufacturing system performance caused by the enlargement of the manufacturing system boundaries, capacities and flexibilities.

Surely one of the most critical benefits achieved by using simulation concerns production planning and control (PPC) aspects. The cooperation among manufacturing firms leads to modify almost every activity of PPC. The concern on internal production planning is replaced by the complexity of external supply planning since this supports the network operation. As soon as a manufacturing unit tries to achieve coordination with its partners, it quickly faces difficulties associated with different operational conventions, locally specific constraints, software legacy and properties, conflicting objectives and misaligned incentives.

In particular, PPC in a production network concerns the coordination of production planning processes across multiple facilities at different hierarchical levels. In Figure 8-1, starting from the upper-left box and proceeding clockwise, three main levels of production planning processes can be distinguished. For each level, a simulation model could be required for testing operation solutions, and for aiding the planner in every phase of these processes.

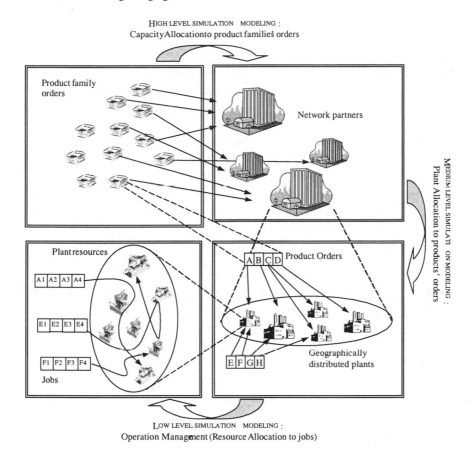

Figure 8-1. Simulation modeling levels for PPC in production networks

The first level, the high level (the upper boxes in the figure), concerns the allocation of families of orders to members of the network. This choice depends on considerations regarding the role of the member in the network, its core competencies, its manufacturing capacity shared with the network, and so forth. Then, the orders have to be allocated to specific plants of the firm, which can be geographically dispersed. This is considered the medium level production planning process (the right boxes in the figure). In this complex task, it must be taken into consideration that, for instance, each order has its expected processing time and each plant has a specific ability to process orders more or less fast, depending on the manufacturing and technical characteristics of the plant, on the status of its resources, and on the specific order under consideration. Lo Nigro et al. (2002) developed a negotiation protocol and a distributed framework based on autonomous agents for the capacity allocation process in a primary electronic company whose plants are distributed geographically worldwide. Once again, they used simulation, by means of a proper simulator in order to test the paybacks and drawbacks of their approach.

Finally, going down to the low level of the production planning process, single jobs need to be allocated to specific manufacturing resources (bottom boxes in the figure). In this case, operation management issues, such as dynamic scheduling decisions or error-handling policies need main attention, and discrete event simulation as a tool has been widely recognized as one of the most powerful and effective tool for solving operational problems in a dynamic manufacturing environment in real-time. Ordinary scheduling systems, dynamic scheduling systems, error handling management policies, task planning issues, etc. are examples of operational issues often investigated by using discrete event simulation modeling tools (Renna et al. 2001, Beltz and Mertens 1996, Heng et al. 2000).

3. MAS FRAMEWORK AND SIMULATION MODEL FOR HIGH LEVEL COORDINATION IN PN

In this section, the author wants to show how a very simple multi agent framework could be developed for studying the problem of the allocation of customer orders to members of a production network. This activity is quite complex for many reasons, as described in previous sections of this Chapter. Roles of partners within the network, their competencies, skill, production capacity, geographical distance from the customer, and other parameters need to be taken into consideration. Also, constraints and variables depending on the order itself play an important role in the decision. Technological requirements of the order, its size, and due date cannot be disregarded. Furthermore, this section presents the simulation model developed for analyzing and testing the above mentioned decentralized decision-making process.

3.1   The MAS model

The production planning control system is based on an agent-based approach that models a set of enterprises, members of the same production network and a set of production orders to be completed.

The architecture of the proposed control system consists of a set of Enterprise Agents, associated to each firm, which have a production capacity, a certain competence, and a specific skill, and a set of Order Agents, associated to each order introduced into the system.

The allocation of orders to members of the production network is accomplished by means of a negotiation between the Order Agent associated to the order, and Enterprise Agents associated to the members. Figure 8-2 portrays the system architecture.

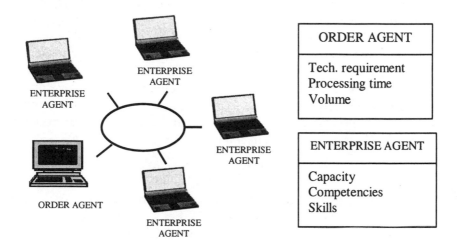

ORDER AGENT

Tech. requirement
Processing time
Volume

ENTERPRISE AGENT

Capacity
Competencies
Skills

Figure8-2. MAS Architecture

The negotiation process starts each time a new order enters the system. When this happens, a network member needs to be allocated to the order; the negotiation among Order Agents and the Enterprise Agents is obtained through a contract net. Briefly, the Order Agent announces the availability of an order to be processed to all Enterprise Agents; who evaluate the announcement and, if they possess the competency to process that order provide an evaluation in processing that order to the respective network partner. The Order Agent merges the evaluations and based on the PN status decides on the performance measure for each partner. The Order Agent then selects the enterprise with the maximum performance. Specifically, the Order Agent calculates the enterprise performance on the basis of the resource skill (i.e., the ability to process that order) and on its total workload (in process, and in queue).

Thus, given a production network consisting of E members (e = 1, 2, ...E), and a set of O types of orders (o = 1, 2, ...O), each with a specific technological requirement, a binary variable $x_{e,o}$ is defined. If its value is equal to 0, the enterprise e does not have the competency to process the order o; 1 otherwise. Also, the network member e has a specific skill $s_{e,o}$ in processing the order o. It is assumed that $s_{e,o} \in [0,1]$. If the potential processing time of the order o is t_o, then the actual processing time of that order in the enterprise e is given by equation (1).

$$t_{e,o} = \frac{t_o}{s_{e,o}}$$ (1)

By analyzing every order already assigned to the enterprise (order currently in process, plus orders in queue), the Enterprise Agent can calculate the current total workload when negotiation for the next order takes place.

Actually the total workload measure WL_e^{norm} that the enterprise calculates and sends to the Order Agent is the inverse normalized measure of the total workload, which is mapped into [0,1]. Specifically, if WL_e is the total workload of the enterprise e, WL_{min} and WL_{max} are respectively the min and max values of the current workloads of all the enterprises, the normalized measure WL_e^{norm} is calculated as in equation (2)

$$WL_e^{norm} = \frac{WL_{max} - WL_e}{WL_{max} - WL_{min}}$$ (2)

The enterprise performance measure in processing the order o, calculated by the Order Agent is·given by equation (3). For sake of clarity, performance of the enterprise utilizing *management policy 1* (MP1) is indicated as:

$$P_{e,o} = s_{e,o} + WL_e^{norm}$$ (3)

3.2   The simulation model

A discrete event simulation utilizes Arena software (Rockwell Software, Inc.) was developed to implement the PN architecture and negotiation mechanism described in the previous Section. Given this simple architecture and negotiation mechanism, the aim at this point is to build up a simulation model of the system and the author selected the Arena® discrete event simulation platform by Rockwell Software, Inc., for such a task.

Discrete event simulation – in many commercial tools and simulation packages, nowadays the simulation model is automatically created from high level modeling languages and notations – allows to validate and optimize dynamic and discrete systems such as production systems, but also workflows such as negotiation mechanisms. These models facilitate evaluating different coordination scenarios and maximizing their potential output and benefits. Arena® – based on the well-known SIMAN simulation language - is well suited for modeling shop floors of production systems in which each entity (part) follows a manufacturing route through production resources (servers, material handling systems, buffers, and so forth) (Kelton et al. 1998). For the application described in this Section, instead, a part type is associated to a production order and a production server is associated with a production enterprise. In other words, the Arena® model, which usually represents a production shop floor, reproduces a production network.

The simulation model, therefore, consists of servers (representing the enterprises of the network), circulating entities (representing the orders), and a number of modeling blocks representing Enterprise Agents and the Order Agent. In Figure 8-3, a sketch of the simulation model built into the Arena® environment for a four-member production network is shown. In the sketch, Enterprise Agents and the Order Agent can be identified; the reader can notice the SIMAN blocks "Signal" and "Wait" that model "control" commands to send and wait for signals among agents during the negotiation process.

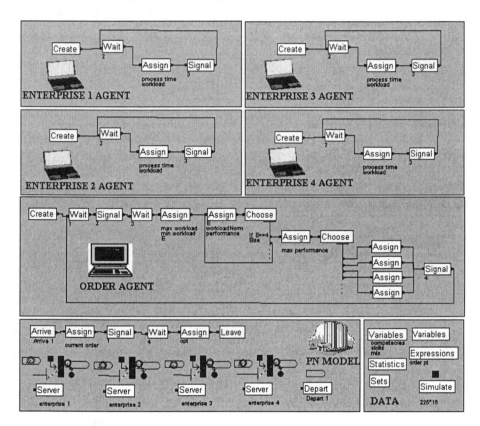

Figure 8-3. The PN simulation model into the Arena environment

The average fulfillment time $AvgF$ has been considered as output performance measure of the coordination strategy and, in particular, of the negotiation process (by following once again the analogy of a production shop floor). This is the average among the times that the entire production network needs to fulfill the order. Such a performance measure has been chosen because of its importance related to a mass customization environment. Indeed, given that the production network has the competency for fulfilling the order launched by the customer (the hypothesis is that there exists at least one member of the network capable of satisfying technological requirements of the order), the most important requirement become the reactivity of the network in releasing the customer order, i.e. to minimize the delivery time.

The simulation model has also been used for testing the management policy (MP1) implemented for the considered order allocation problem and to compare it with a new management policy (MP2). According to this new policy, the performance of each enterprise is calculated only by considering orders currently assigned to the enterprise. Although, this neglects the ability of the member in processing the order during the negotiation process (given by its skill), it clearly overcomes problems arising due to the fact that network members are enforced to communicate their skill in fulfilling the given order. Instead of equation (3), the Order Agent calculates the enterprise performance according to equation (4).

$$P_{e,o} = WL_e^{norm} \tag{4}$$

The next section describes the numerical case study that has been considered as an example.

4 SAMPLE CASE STUDY

A production network of 4 partners has been considered. The network is required to process 5 types of orders characterized by 5 different technological requirements. The order inter-arrival times follow a statistical exponential distribution with mean 1 day (SCENARIO 1), while their processing times and the arrival mix are reported in Table 8-1.

Tables 2 and 3 report enterprise competences and skills, respectively. For example, enterprise $e=2$ has the competency to process orders of kind $o=2$ with a skill equal to $s_{e,o}=0.78$; enterprise $e=1$ can process the same kind of orders with a higher skill equal to $s_{e,o}=0.78$; enterprise $e=3$ does not have the competency to process that order.

Table 8-1. Order Data

O	Order processing times (hours)	mix
1	Triangular (4,5,6)	10%
2	Triangular (5,6,7)	40%
3	Triangular (6,7,8)	10%
4	Triangular (7,8,9)	30%
5	Triangular (8,9,10)	10%

Table 8-2. Enterprise competencies ($x_{e,o}$)

	e=1	e=2	e=3	e=4
o=1	1	1	0	1
o=2	1	1	0	1
o=3	0	1	1	0
o=4	1	1	1	1
o=5	1	1	1	1

Table 8-3. Enterprise skills ($s_{e,o}$)

	e=1	e=2	e=3	e=4
o=1	0.80	0.70	-	0.50
o=2	0.75	0.78	-	0.40
o=3	-	0.70	0.92	-
o=4	0.85	0.65	0.50	0.90
o=5	0.80	0.60	0.80	0.60

The system has been simulated for a year, 225 working days/year, 2 shifts/day, 8 hours/shift. Because of the random inputs of the simulation model, a certain number of replications have to be run in order to guarantee that for the output *AvgF*, the length of confidence intervals (95% level) of the mean among replications is lower than 10% of the mean. Simulation results are reported in Table 8-4.

Table 8-4. Simulation Results – SCENARIO 1(high congestion)

	o=1	o=2	o=3	o=4	o=5
AvgF (MP1)	11.62	13.20	14.48	15.97	17.49
AvgF (MP2)	8.41	9.58	10.45	11.43	12.56

From Table 8-4, it can be inferred that MP2 has to be preferred to MP1, due to lower mean times required to fulfill every type of order. For example, it takes 11.62 hours for fulfilling the order o=1 when MP1 is adopted; and 8.41 hours when MP2 is applied. This means that, even though the adoption of MP2 does not give the Order Agent any information regarding network members' skills for his final choice (see equation (4)), the negotiation mechanism brings better results than those in the case of sharing that information.

The expected order inter-arrival time is now assumed as 2 days (SCENARIO 2), instead of 1 day. In this manner, the traffic of orders into the system is surely less congested and the network partners are less overwhelmed because, on an average, there are less orders circulating in the PN. Results of this condition are reported in Table 8-5.

Table 8-5. Simulation Results – SCENARIO 2 (low congestion)

	$o=1$	$o=2$	$o=3$	$o=4$	$o=5$
AvgF (MP1)	6.86	8.29	8.96	10.03	11.26
AvgF (MP2)	8.00	9.21	10.32	11.89	12.19

In this case, as can be observed, conclusions are totally different with respect to the first scenario. Indeed, when policy MP1 is employed, the average time for fulfilling each type of order is lower than when MP2 is adopted. For example, 6.86 hours for fulfilling order $o=1$ for MP1, and 8.00 hours for MP2. This further result was quite expectable. In fact, in policy MP1, entering orders are dispatched to the network members by means of a negotiation that takes into account (with equal importance), both the current workload and the skill of the enterprises (see equation (3)). If policy MP2 is used, instead, only the current workload is considered (see equation (4)).

Therefore, when the enterprises are congested (scenario 1), it would be more preferable that the member to be selected for allocating the new order is the one with minimum workload, instead of on its skill level (or at least would be better to consider its skill, but with a minor weight than the workload, in making the decision). Therefore, the computation of the member performance given by equation (3) could be generalized as in equation (5) by adding weights α and β:

$$P_{e,o} = \alpha \cdot s_{e,o} + \beta \cdot WL_e^{norm} \tag{5}$$

This generalization allows the proper tuning of the importance of both the workload and the skill of the member for the final decision of order allocation decision. Accordingly, the Order Agent could make its final decision depending on the actual status of the PN by changing dynamically the coefficient α and β. Specifically, the more crowded the network, lower should be the α and higher the β. However, this, while improving the total system efficiency, would require a more substantial information sharing.

5. CONCLUSIONS

During the last decades, manufacturing firms have experienced increasing pressure to improve production efficiency, responsiveness to market changes, and substantial cost reduction. Mass customization and the need to be part of volatile businesses are, nowadays, faced by two major paradigms: reconfigurable manufacturing and reconfigurable enterprise. Publicized research and industrial applications show that reconfigurable manufacturing systems and production networks are the natural response to such paradigms, being key-enablers for mass customizable production.

This Chapter focuses on production networks and highlights their impact on mass customization. Then, it puts into evidence how the design, management, and coordination of such dynamic and complex systems, are crucial tasks and require preliminary analysis phases before their implementation. Agent based frameworks and simulation modeling are proven to be properly suited in approaching these problems.

Examples of use and application of these engineering tools in the design and management of production networks have been given in the Chapter. Particular attention has been given to the description of how simulation can be utilized in conjunction with multi agent system technologies to develop realistic prototypes of decision support systems for complex problem solving in large-scale production network.

A sample agent based framework for production network has been proposed and an appropriate simulation model has been demonstrated to support the decision making for the specific problem of network partners allocation to customer orders. The presented numerical example shows a potential utilization of the simulation tool. In particular, two management policies associated with the agent negotiation mechanisms have been compared in two different market scenarios related to different levels of congestion and traffic of order circulating into the network.

The author believes that, given the importance of production network as the most practiced industrial response to mass customization, several research challenges arise. One of the most important and crucial issue concerns the use of Information and Communication Technologies to build real distributed management tools for production network. In fact, only through proper coordination tools, distributed networks can match reactiveness and efficiency, the two major requirements of mass customization. As the example presented shows, the possibility to build such tools relies on agent-based system technology, which allows the design of proper coordination and negotiation strategies. Once again, simulation can play a very important role on designing such ICT based management tools. Simulation tools, indeed, can be used to develop and test the correct use of such technologies in order to develop proactive and efficient production network management tools.

ACKNOWLEDGEMENTS

The author wishes to thank the editors and the anonymous referee whose comments have greatly improved this Chapter. The author is also grateful to Prof. Giovanni Perrone and Prof. Sergio Noto La Diega for their helpful remarks and comments that made possible the realization of his contribution to this book.

REFERENCES

1. Ahn, H.J., Lee, H., Park, S.J. A flexible agent system for change adaptation in supply chains. Expert Systems with Applications 2003; 25: 603-618.
2. Belz, R., Mertens, P. Combining knowledge-based systems and simulation to solve rescheduling problems. Decision Support Systems 1996; 17: 141-157.
3. Brandolese, A., Brun, A., Portioli-Staudacher, A. A Multi-Agent Approach for the Capacity Allocation Problem. International Journal of Production Economics 2000; 66: 269-285.
4. Bruccoleri, M., Amico, M., And Perrone, G. Distributed intelligent control of exceptions in reconfigurable manufacturing systems. International Journal of Production Research 2003; 41/2: 1393-1412.
5. Bruccoleri, M., Pasek, Z. Operational Issues in Reconfigurable Manufacturing Systems: Exception Handling. Proceedings of the 5th biannual World Automation Congress 2002; 3; June 9-13; Orlando Florida.
6. Cantamessa, M., Fichera, S., Grieco, A., La Commare, U., Perrone, G., Tolio, T. Process and production planning in manufacturing enterprise networks. Proceedings of the 1st CIRP Conference on Digital Enterprise Technology 2002; : 187-190, Durham.
7. Ferber, J. Multi-Agent Systems. An Introduction to Distributed Artificial Intelligence. Addison-Wesley, 1999.
8. Fox, M. S., Barbuceanu, M., Teigen, R. Agent-Oriented Supply Chain Management. The International Journal of Flexible Manufacturing Systems 2000; 12: 165-188
9. Fuji, S., Kaihara, T., Morita, M. A distributed virtual factory in agile manufacturing environment. International Journal of Production Research 2000; 38/17: 4113-4128.
10. Goldhar, J., Jelinek, M. Plans for economics of scope. Harvard Business Review 1983; 61/6: 141–148.
11. Heng, L., Li Z., Li L. X., Bin H. A production rescheduling expert simulation system. European Journal of Operational Research 2000; 124: 283-293.
12. Katzy, B. R., Dissel, M. A toolset for building the virtual enterprise. Journal of Intelligent Manufacturing 2001; 12: 121-131.
13. Kelton, W. D., Sadowsky, R. P., Sadowsky, D. A. Simulation with Arena®. New York: Avenue of Americas, Mc Graw Hill 1998.

14. Koren, Y., Heisel, U., Jovane, F., Moriwaki, T., Pritschow, G., Ulsoy, G., Van Brussel, H. Reconfigurable Manufacturing Systems. Annals of CIRP (Keynote Paper) 1999; 48/2.
15. Lo Nigro, G., Noto La Diega, S., Perrone, G., Renna, P. Coordination policies to support decision making in virtual organization. Proceedings of the International Manufacturing Leaders Forum; 2002; Adelaide Australia, 166-173.
16. Martinez, M. T., Fouletier, P., Park, K. H., Favrel, J. Virtual enterprise – organization, evolution and control. International Journal of Production Economics 2001: 74: 225-238.
17. Pine, B. J., Victor, B., Boyton, A.C., 1993. Making mass customization work, Harvard Business Review 71, 108–119.
18. Renna, P., Perrone, G., Amico, M., Bruccoleri, M, Noto La Diega, S., 2001, A performance comparison between market like and efficiency based approaches in Agent Based Manufacturing environment. Proceedings of the 34th CIRP International Seminar on Manufacturing Systems; 2001 May: 93-98; Athens.
19. Shen, W., Norrie D., H. Agent-Based Systems for Intelligent Manufacturing: A State-of-the-Art Survey. Knowledge and Information System, an International Journal 1999; 1/2: 129-156.
20. Smith, R. G. The contract net protocol: high level communication and control in a distributed problem solver. IEEE Transactions On Computers 1980: 12: 1104-1113.
21. Swaminathan, J. M., Smith S. F., Sadeh, N. M. Modeling Supply Chain Dynamics: A Multiagent Approach. Decision Sciences 1998; 29/3: 607-632.
22. Turowski, K., 2002, Agent-based e-commerce in case of mass customization, International Journal of Production Economics, 75, 69-81.
23. U.S. National Research Council, 1998, Visionary Manufacturing Challenges for 2020, National Academy Press, Washington, Dc, U.S.A.
24. Virtual Manufacturing Technical Workshop - Dayton, Ohio 25-26 October 1994, Technical Report - March 07, 2001.
25. Wiendhal, H. P., Lutz, S. Production in Networks. Annals of CIRP 2002; 51/2.

CHAPTER 9

KNOWLEDGE MANAGEMENT FOR CONSUMER-FOCUSED PRODUCT DESIGN

Charu Chandra[1], Ali K. Kamrani[2]

[1]University of Michigan-Dearborn
[2] *University of Houston*

Abstract: As firms adopt a consumer focus for mass customizable product development strategy, it becomes essential for them to conduct early product design and development trade-off analysis among competing objectives of increased product variety, shorter product lifecycles, and smaller lot sizes. A distributed Knowledge Base System is needed for these complex decisions. This chapter proposes a knowledge management approach based on consumer-focused product design philosophy. It integrates capabilities for (a) intelligent information support, and (b) group decision-making, utilizing a common enterprise network model and knowledge interface through shared ontologies.

Keywords: Mass customizable product development, knowledge management, ontology based supply chain modeling.

1. INTRODUCTION

As consumer focus increasingly drives the product development process, manufacturers are adopting the strategy of offering enhanced product variety, reduced product time-to-market, and flexible manufacturing to process orders in arbitrary lot sizes. A key facet of this strategy is the ability to conduct trade-off analysis among these competing objectives, on early product design and development. For the delivery of the product, it requires co-ordination of related business *processes* (workflow), spread across locations in the extended supply chain. A distributed Knowledge Management environment comprising, (a) a Knowledge Based System to analyze, verify, store and retrieve process definitions, and (b) a Process Manager, to manage execution of processes defined in the Knowledge Base System, is needed to implement the complexity of workflow. Information technologies are essential to disseminate product knowledge and integrate the decision-making process among heterogeneous and distributed members of the product supply chain.

This Chapter describes a knowledge management approach for a consumer-focused product development philosophy. It proposes integration of intelligent information support and group decision-making capabilities, utilizing a common enterprise network model and knowledge interface through shared ontologies. First, consumer-focused product design for manufacturing is described. Then, we discuss requirements for knowledge management in consumer-focused product development. We describe a framework that meets these requirements. It prescribes shared ontologies for consumer-focused product development, and lays out the blueprint of integrating information support for decision-making activities in an enterprise. It is based on modeling an enterprise as a dynamic constraints network and has three main components, viz., a methodology, techniques, and tools. Finally conclusions and directions for future research in this topic are highlighted.

2. CONSUMER-FOCUSED PRODUCT DESIGN

Manufacturers increasingly target products to meet consumer needs and preferences while exploring newer markets. Products are being designed to offer both tangible and intangible benefits. Maintaining a balance between anticipated product features and benefits is a pre-requisite. A consumer-focused product design strives to simultaneously satisfy some of the conflicting objectives of consumer and manufacturer / seller. It is based on premises that, (a) changing customer requirements dictate varied product features, (b) structure of products and processes must be aligned with dynamic product features, and (c) manufacturing productivity requires managing conflicting objectives due to these structural alignments. We discuss each of these below.

- The first premise ensures that product features are designed to offer, (a) style and technology to satisfy technical feasibility, (b) utility, value, and price to meet economic feasibility, and (c) quality, and reliability to meet operational feasibility of product design.
- The second premise mandates clustering products based on common product features (attributes) and then mapping these to identical processes and / or operations. This strategy results in reduced – lead-time, set up time, resource utilization, process flow, and costs.
- The third premise requires managing effects of first two premises on manufacturing productivity in the presence of multi objectives caused by product-process realignment. That is, the impact of product variety management on time-to-market (lead-time, set up time), cost, and scaling of manufacturing / production operations (lot sizing).

2.1 Trends in Consumer Focused Product Design

The trends related to consumer-focused product development are: product-process orientation, enterprise reconfiguration, and the use of information technologies in decision-making. Some of these trends are described below.

2.1.1 Mass Customization and Product Postponement

Mass customization is achieved through product configuration so as to adapt to customer requirements by mass production of individually customized goods and services. Da Silveira et al. (2001) argue that a balance between customer expectations about price and delivery promptness of customized products, and producers' ability to meet them within acceptable cost and time frame is one of the conditions required for successful implementation of mass customization. They also point out that an efficient production network should be available for implementation of this strategy. If other conditions are met, then finding the balance and a production network configuration supporting it remains an ultimate decision making goal. The balance can be characterized by delivery time and inventory held to secure the service level. In such a setting, the problem of adopting mass customization policies directly ties with postponement. Customization is achieved by postponing some production activities until customer orders are received. Production is finalized according to ordered product specification. In a multi-stage production system, postponement would signify a stage, starting from which production is commenced only upon receiving customer's order. Postponement may entail re-designing products using modularity and commonality as design principles (Van Hoek 2001). Also influenced are product and process design so as to differentiate between discrete, continuous, and decoupled processes.

2.1.2   Mass Customization and Supply Chain

Implementation of mass customization and postponement strategies affects the enterprise structure because postponement activities will most likely be placed close to the market. As products are designed in accordance with these strategies, it is imperative their effects be implicitly reflected in the design of supply chain from sourcing to final distribution of products. Supply chain is a special form of complex business enterprise system, where the key is to co-ordinate information and material flows, plant operations, and logistics (Lee and Billington 1993). The fundamental premise of a supply chain management approach is synchronization among multiple autonomous business entities represented by it. That is, improved co-ordination *within* and *between* various supply chain members. Co-ordination is achieved within the framework of commitments made by Members to each other. Increased co-ordination can lead to reduction in lead times and costs, alignment of interdependent decision-making processes, and improvement in the overall performance of each Member, as well as the supply chain network (Group) (Chandra 1997, Sousa et al. 1999). The most common form of supply chain decision-making is aimed at managing business-to-business, and business-to-consumer model for service and goods transactions.

2.1.3   Mass Customization and Information Technologies

Information technologies designed and implemented for business-to-business, and business-to-consumer models impact the market substantially by driving costs down through standardized networking technologies, and creation of entirely new enterprise and/or relationships by real-time interconnection of companies with their customers. These technologies have a strategic objective of managing customers' needs by way of a proactive "consumer pull", as against the traditional "product push" strategy. A heavy emphasis is placed on customer relationship management, which involves identifying *goals* (customers wants and needs), and developing marketing programs aimed directly at fulfilling these goals.

Implementation of above information technology strategies requires capabilities for real-time decision support. The challenge is to accommodate interaction of various units with competing objectives as they share information with each other towards achieving their shared goals. Therefore, e-management has come to symbolize a management philosophy that reflects important traits of the global digital economy, namely dynamic real-time decision-making, customer orientation, and speed in responding to market demands. This topic has evinced interest on various aspects of the problem based on Internet technologies, such as e-commerce, e-business, and e-manufacturing.

2.2   Impact on manufacturing operations

An intelligent enterprise, such as a supply chain with e-management capabilities may be realized in the form of a virtual enterprise supporting manufacturing operations. The major component of this enterprise is intelligent information support utilizing advanced information technologies. Decision-making capabilities in this enterprise are dynamic. Operations of such enterprise are also labeled as virtual manufacturing, and / or e-manufacturing.

Virtual enterprise is a temporal co-operation of independent units, which provide a service on the basis of shared skills, technologies and resources. Its components are a set of technology and associated resources available to a unit(s) belonging to a virtual enterprise at various time intervals. Units of a virtual enterprise may share common technology and associated resources through coalition agreement(s). Sharing is facilitated through logistic systems, connected via common objectives and policies for implementation.

One of the phases of virtual enterprise creation is its configuration, necessitated by market conditions that dictate uniquely structuring its product, process, and resource components. Interaction among these components brings complexity to enterprise structures. To deal with such a complexity, an enterprise needs knowledge management and an integrated knowledge base system for its support. Therefore, knowledge about an enterprise must be designed with the view of integrating its product, process, and resource components. One of the emerging forms of a virtual enterprise is a supply chain, the basis of knowledge management framework described in the rest of the chapter.

3.   REQUIREMENTS FOR KNOWLEDGE MANAGEMENT IN CONSUMER-FOCUSED PRODUCT DESIGN

For efficient management of supply chain, it is essential that its design be properly configured. An important facet of configuration is to understand the relationship between supply chain system components that define its structure, namely products, processes and resources. The primary goal is to facilitate transfer and sharing of knowledge in the context of new forms of supply chain configurations, developed in response to changing consumer focus in product development.

3.1   Requirements for knowledge integration in decision-making tasks

Requirements for knowledge integration in an intelligent enterprise must reflect interconnectedness between problem solving approaches and technologies for the enterprise information environment. The proposed approaches and associated technologies are categorized into two groups, (i) problem solving, and (ii) information support. For the first group, these are (1) mass customization management, (2) configuration management, and (3) constraint satisfaction and propagation. For the second group, these are (a) data and knowledge management, (b) ontological representation of knowledge, (c) multi-agent and intelligent agent, and (d) conceptual and information modeling. For the purpose of the framework described in this chapter, we focus our attention on items (a) and (b) of the second group with their interconnections to remaining items of both groups, respectively.

Configuration Management: The objective of designing an intelligent enterprise utilizing configuration principles is to generate customized solutions based on standard components, such as templates, baselines, and models. There are two aspects to configuration management, (i) configuring / reconfiguring, and (ii) configuration maintenance. Configuring deals with creating configuration solutions and selecting components and ways to configure these. Configuration maintenance deals with maintaining consistency among selected components and decisions.

Constraint Satisfaction and Propagation: In the design of a supply chain, it is often necessary to solve a dynamic constraint satisfaction problem, where applicable constraints depend upon various design aspects and time horizons.

Multi-agent and Intelligent Agent: Implementation of the basic principle of cooperation in the supply chain is based on distribution of procedures between different units / users (or agents) concurrently in the common knowledge space. It is, therefore, natural to represent configuration management knowledge as a set of interacting autonomous agents in a multi agent environment. Agent is a mechanism that facilitates capturing behavioral characteristics of the problem for a specific process or activity. Intelligent agent is an entity that can navigate in heterogeneous decision-making environment, and either alone or working with other agents, achieve specific goal(s).

Conceptual and Information modeling enables representation and evaluation of system entity characteristics, relationships to other entities, and controls to achieve objectives. Some of the techniques utilized for evaluation of various enterprise configurations are: process modeling, entity relationship modeling, object-oriented modeling, and genetic algorithms.

In order to coordinate flow of materials within a multi-echelon supply chain, it is important to synchronize activities both at inter and intra levels by sharing information. To support this objective, an information kernel in the form of a *"supply chain conceptual model"* is needed. This kernel describes major components of a supply chain as a set of: objectives at strategic level, viz., supply chain model attributes, strategies, supply chain units, constraints for every unit, products for every unit, unit resources, contract relationships among units, and coefficients for bilateral relationships among units.

Data and Knowledge Management offers intelligent support, critical to realizing competitive advantage for a supply chain. System information integration deals with achieving common interface "within" and "between" various components at different levels of hierarchies, different architectures and methodologies (Hirsch 1995, Sousa et al. 1999) using distributed artificial intelligence and intelligent agents (Fischer et al. 1996, Gasser 1991; Jennings et al. 1996, Jennings 1994, Lesser 1999, Sandholm 1998, Smirnov 1999, and Wooldridge and Jennings 1995). Knowledge modeling and ontologies describe unique system descriptions that are relevant to specific application domains (Fikes and Farquhar 1999, Gruber 1995, Gruninger 1997, Smirnov and Chandra 2000, Chandra et al. 2000).

Knowledge is a set of relations (constraints, functions, and rules) by which a user (agent) decides to use the information in order to perform timely actions in meeting a goal or a set of goals. Knowledge is key to managing collaborative activities in the supply chain. Therefore, knowledge must be relevant to overall business goals and processes and be accessible in the right form when needed. An important requirement for an integrated Knowledge Base System is the ability to capture knowledge from multiple domains and store it in a form that facilitates reuse and sharing. This is accomplished via design and development of Knowledge Management mechanisms with following knowledge levels:

- System knowledge describing rules for integration between units (manufacturer and its extended supplier and dealer network) of the enterprise, and its management and maintenance.
- Facilitator knowledge describing rules for distribution of knowledge and identification of access level in sharing data and knowledge base.
- Unit knowledge describing reusable methods, techniques and solutions for problem solving at the unit level.

- User knowledge describing knowledge related to individualized special skills of a user at the problem domain level.

Ontology is a form of knowledge representation applied to various domains. A FIPA definition of ontology describes it as an explicit specification of the structure of a certain domain (FIPA 1998). Ontology includes a vocabulary (i.e., a list of logical constants and predicate symbols) for referring to the subject area, and a set of logical statements expressing constraints existing in the domain and restricting the interpretation of the vocabulary. Ontology provides a vocabulary for representing and communicating knowledge about some topic and a set of relationships and properties that hold for entities denoted by that vocabulary.

Ontology is useful in creating unique models of a supply chain by developing knowledge bases specific to various problem domains. Ontologies are managed by translation and mapping between different types of entities and attributes. These capture rules and constraints for the domain of interest, allowing useful inferences to be drawn for execution, analysis, and validation of models. Ontological translation of an enterprise, such as a supply chain is necessary because networks are multi-ontology classes of entities. Various ontologies for an entity describe its unique characteristics in context with the relationship acquired for a specific purpose or problem. Each user (agent) works with its own ontology-based knowledge domain model (Fikes and Farquhar 1999, Jennings 2000, Fox and Gruninger 1999, Uschold and Gruninger 1996, Ruberstein-Montano et al. 2001, and Vasconcelos et al. 2000).

4. A KNOWLEDGE INTEGRATION FRAMEWORK FOR CONSUMER-FOCUSED PRODUCT DESIGN

A framework of knowledge management for consumer-focused product design that integrates information support for decision-making activities in an enterprise is presented in this section. It is based on modeling an enterprise as a dynamic constraints network and has three main components:

1. A methodology for supporting knowledge management needs of a dynamic constraints network,
2. Techniques for design and modeling of knowledge base system based on taxonomical descriptions of the problem-solving environment in a complex enterprise system, and
3. Tools for design, modeling, implementation and evaluation of the problem-solving environment in a complex enterprise system.

Each of these components is described below in the context of a supply chain.

4.1 Methodology:

In order to design a supply chain that can be configured to meet changing production needs, relationship between system structures due to "product–process–resource" interactions must be understood. Supply chain configuration generates customized solutions based on standard components (as templates or baselines), or supply chain model. Hence, the implementation of supply chain approach is based on the shareable information environment that supports the "product-process-resource" model of an enterprise. A generic enterprise model may be defined as follows:

- Natural language explanation of the meaning of modeling concepts -- *glossaries of terms.*
- Some forms of Meta models, e.g., *process models, entity relationship, Meta schema, and conceptual models of terminology component of modeling languages*, describing relationships among modeling concepts.
- Ontological theories defining the meaning (*semantics*) of enterprise modeling concepts, in order to improve the analytic capability of decision-making, and through these the usefulness of enterprise models.

An important requirement for a collaborative system is the ability to capture knowledge from multiple domains and store it in a form that facilitates reuse and sharing (Sousa et al. 1999). The methodology suggested in this chapter is limited to designing knowledge management capabilities for product-process-resource configurations, focused on utilizing *reusable knowledge* through ontological descriptions of a dynamic constraints network. This is accomplished by knowledge modeling product, process, and resource components to satisfy manufacturing constraints in a firm's environment. Reusable knowledge management deals with organizing "knowledge clusters" by their inherently common characteristics as observed in various problem domains; and utilizing these as templates to describe unique conceptual models of an enterprise or its components. It is based on GERAM, the Generalized Enterprise Reference Architecture and Methodology (ISO TC 184/SC 5/WG 1, 1997) at the domain level; and MES (MESA 1998), MEIP (MEIP 1999), NIIIP (NIIIP 1994) and WFM (WFM 1996) methodologies at the application level.

A dynamic constraints network provides the basic structure for supply chain configuration in the above methodology. Constraint satisfaction is a fundamental problem for solving supply chain issues. Conventional constraint satisfaction procedures are designed for the problem with one constant set of constraints. However, in manufacturing systems (design for productivity, configuration, layout, and scheduling), it is often necessary to solve a dynamic constraint satisfaction problem where applicable constraints depend on design aspects (Smirnov 1994). The domain knowledge model of supply chain contains entities (objects), which can be of different types (classes). Multi-level representations are used for product-process-resource model description. In addition, each unit (supply chain Member) may work with its own ontology-oriented constraint network. Thus, the dynamic constraints network approach presented in this chapter with interconnectivity between design and production features reflecting complete association of product, process, and resource components in the supply chain offers the capability to model integrated solutions to problems.

An abstract product-process-resource model is based on the concept of ontology-oriented constraint networks. Multi-ontology classes of entities, attribute logic and the constraint satisfaction problem model represent networks. This abstract model unifies main concepts of languages, such as standard object-oriented languages with classes, and constraint programming languages. It supports the declarative representation, efficiency of dynamic constraint solving, as well as problem modeling capability, maintainability, reusability, and extensibility of the object-oriented technology (Smirnov and Chandra 2000).

The ontology-oriented constraints network model is denoted, A = *(St, C)*, where *St* — is an ontology structure, *C* — is a set of ontology constraints. To deal with the conceptual schema of configuring process defined in terms of constraints, a dynamic constraints network model is applied. A static constraints network $A_i = (V_i, D_i, C_i)$, involves a set of variables $V_i = (v_{i1}, v_{i2}, \ldots, v_{iN_i})$, each taking value in its respective domain $D_i = D_{i1} \times D_{i2} \times \ldots \times D_{ij} \times \ldots \times D_{iN_i} = \underset{j=1}{\overset{N_i}{\times}} D_{ij}$, and a set of constraints $C_i = \{c_{i1}, c_{i2}, \ldots, c_{ik}\}$. A dynamic constraint network N is a sequence of static constraints networks, each resulting from a change in preceding one imposed by the external environment.

For design of supply chain knowledge base, we utilize ontology design. It is based on an ontology hierarchy, depicted in figure 9-1. The top-level ontology is the "shared ontology" for domain independent representation of the problem set. This type of ontology is needed to describe an abstract model using common knowledge representation. The lower-level ontology is "application ontology" and is a combination of the "domain specific ontology" and the "problem-specific ontology". This type of ontology is needed to describe special knowledge about an application or a problem for unit and user. The top-level ontology is oriented for dynamic constraints network, while the lower-level ontology is for ontology-based constraints network. The product configuration is represented by the following relationship: "*configuration of the product* (product structure, materials bill) → *configuration of the business process* (process structure, operation types) → *configuration of the resource* (structure of system, equipment and skill levels)".

Applying above methodology enables forming the conceptual model of the supply chain system. This is accomplished by knowledge modeling its product, process, and resource components to satisfy manufacturing constraints in its environment. The implementation of this approach is based on the shared information environment that supports the product-process-resource model used for integration and co-ordination of user's (unit's) activity. It identifies relationships between various model types of the dynamic constraints network, their relationships to user types (agents) and respective data and knowledge needs.

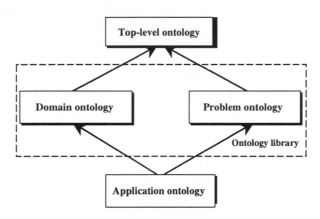

Figure 9-1. An ontology management hierarchy for Knowledge Base System

An *Ontology Management Agent* performs the function of employing the product-process-resource model at various levels of decision-making to provide solutions for problems at varying levels of complexity. For instance, models for product-customer or product-supplier are modeled at higher levels of abstractions since decision-making is at strategic level for complex top-level business problems. However, product-process-resource model for logistic manager are modeled at detailed levels for low-level routing, invoicing, and packaging problems. Accordingly, the system is supported by a number of ontologies, which are deployed as the decision-making process is invoked at various levels:

- *Domain Ontologies* are designed to provide common high-level knowledge related to system structures and controls. The product, process, and resource system knowledge, their interactions in formulating various supply chain structures and controls that set boundaries of these relationships are examples of this type of ontology. It is usually designed for industry specific supply chain, or important function modules thereof.

- *Service Ontologies* are designed to provide low-level knowledge needed to perform service functions for a larger strategic problem by solving lower level problems that improve operational productivity in a supply chain. For example, for a strategic production planning and control problem in a supply chain, service ontologies will offer knowledge for specific solutions to demand forecasting, inventory management, capacity and production planning for multi-echelon planning systems.

- *Ontology of administration and management* are designed to provide knowledge to facilitate implementation of technical tasks, such as communication amongst various supply chain users (agents) utilizing appropriate protocols. It identifies administrative details such as, the unique address, as well as the networking protocol to be used when interfacing with the software system.

- *Ontology of roles* – Denotes roles and terms of engagement for transactions that agents may wish to play, namely supplier, consumer, producer, negotiator, and bidder as they negotiate services in the supply chain.

The above Ontology Management approach is based on two mechanisms (Smirnov and Chandra 2000): (1) object class inheritance mechanism supported by inheritance of class ontologies (attributes inheritance) and by inheritance of constraints on class attribute values, and (2) constraint inheritance mechanism for inter-ontology conversion supported by constraint inheritance for general model (constraints strengthening for "top-down" or "begin-end" processes).

4.2 Implementation Techniques

For implementation of the methodology described in Section 4.1, a conceptual framework for supply chain information systems support architecture depicted in figure 9-2, is proposed using system taxonomy, process models, and ontologies. First, we offer below an overview of the framework, followed by description of its various elements.

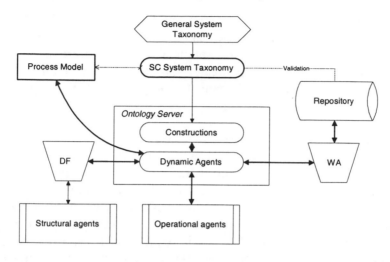

Figure 9-2. Supply chain of information system support conceptual framework

General system taxonomy presents system components at highly generalized level. This presentation can be applied to any type of system. Supply chain inherits its features, thus offering system view of supply chain components and activities. Process model constructions are concerned with supply chain problem solving, modeling, analysis and implementation on the basis of supply chain problems taxonomy. The general purpose is to ascertain how characteristics of the product, process, and resource elements of the supply chain depend on the specific problem environment and to elaborate various supply chain process models. According to system approach applied to ontology development, ontological constructions must be based on system taxonomy. Supply chain system taxonomy is a class structure, where supply chain characteristics are presented comprehensively. Ontological constructions are created from taxonomy, copying relationships between its characteristics and groupings. Every ontology is a subunit of taxonomy. Besides ontology structures must be validated by taxonomy, providing an overseeing framework to ensure that general requirements are addressed.

Ontology server is a combination of ontological construction and dynamic agent modules. Dynamic agents are not real agents. They are mechanisms for creating ontologies as knowledge modules from ontological construction, which are actually knowledge representation formats only. Dynamic agents populate ontological constructions with data taken from central repository, such as an Enterprise Resource Planning System database and send these to operational agents, which are problem-solving modules. Ontologies are also utilized by Structural agents, which are supply chain members, or groups of members. Dynamic agents also provide connection between Process Model structures and ontology, thus updating knowledge acquired from process models to central repository, or to operational agents.

Ontology is a structured and explicit object-oriented tree representation of characteristics about a particular problem-solving environment, or information about a specific domain. Ontologies are distinguished by two distinct ontology-types:

- Domain ontology
- Problem solving or service ontology

Ontologies may be represented as a scheme in XML files, which are supported by a majority of platforms and software development tools. The purpose for building ontology server is to enable technology that will provide large-scale reuse of ontologies, not only inside the enterprise, but also at a distributed level. Before building the server, ontologies must be built. Building ontology for a particular domain requires analysis, revealing relevant concepts, attribute relations, and constraint of the domain. This knowledge is acquired from taxonomy.

4.2.1 General System Taxonomy

General system taxonomy seeks to incorporate process, environmental and other variables in a system. The resulting system is based on variables that are measurable and tractable (Bertalanffy 1975, Lambert and Cooper 2000, McCarthy 1995, McCarthy and Ridgway 2000, and McKelvey 1982). The complexity and dynamic nature of contemporary manufacturing organizations complicate understanding about them. Separating system components with underlying variables into modules helps to map the system for further modeling and analysis. The proposed system hierarchy separates system taxonomy into three levels: System, Enterprise System, and Supply Chain System. For each level, a class diagram is proposed according to object-oriented modeling techniques.

Object-oriented approach attempts to create a hierarchy of classes. The most general class includes parameters and procedures that are relevant to any system. It can be an industry such as automotive, textile, or a small manufacturing line. The top class in our structure is system in general. System scientists describe general system with seven aspects, depicted in figure 9-3, part of a super system. The hierarchical structure is based upon the need for more inclusive clustering or combination of subsystems into a broader system, in order to coordinate activities and processes.

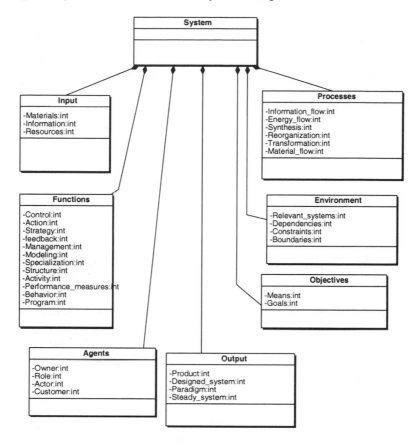

Figure 9-3. General System Taxonomy

4.2.2 Supply Chain System Taxonomy

The system taxonomy is developed through systematic analysis of supply chain and enterprise characteristics. These characteristics illustrate supply chain and enterprise activities and processes. Generalizing from variety and complexity of parameters describing supply chain, essential parameters are chosen. Characteristics of a supply chain are not only distinguished by physical connections (number of products, types of participants, etc.), but also by operations, objectives and attributes such as manufacturing processes, business objectives and inventory needs. First a set of variables are defined and labeled operational taxonomic units, corresponding to general system components. The approach employed, first, reduces the concentration of data, accomplished by packing data in smaller groups of variables. Then, the configuration approach is used to identify classification or taxonomy of overall system. Supply chain, assumed as being a specialization of general system, inherits its structure and parameters. The final structure depicted in figure 9-4, represents general and special components distributed in seven subunits (adopted from general system taxonomy). In these subunits, characteristics are represented by small groups of attributes. What differentiates supply chain from general system is its specialization level, whereby; all components are more specific and relevant to a supply chain domain.

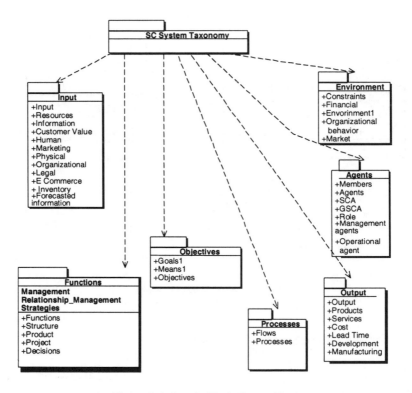

Figure 9-4. Supply Chain System Taxonomy

The organization of proposed taxonomy was designed to accommodate supply chain and enterprise characteristics, which will help to solve supply chain management problems, thus providing information support system with systematic mechanisms for dealing with complex data. The supply chain system taxonomy proposed in this chapter has aided in the development of system representation of supply chain, with whole-part relationships. It utilizes an object-oriented representation, focused on effective modeling of information system, dealing with tasks of supply / production / distribution.

Utilizing these characteristics, we can start building information meta-model by:

- Starting with system taxonomy structure,
- Finding classes, where above mentioned characteristics exist and selecting these for the problem, and
- Deleting all other classes and packages.

As a result, an information presentation format is developed, which can be used by the model. There are two main reasons for utilizing these steps, instead of using the plain list of characteristics. First, these steps offer a standardized format needed by computational tools, which is common for every problem-solving environment. Second, these steps allow reusability of the same structure, whereby any other module can use results from this problem-solving module. Information structure meta-model for Inventory Control utilizing this technique is depicted in figure 9-5.

4.2.3 Product-Process-Resource Model Taxonomy

The supply chain configuration is presented by following relationship: "*configuring the product* (product structure, materials bill) -> *configuring the business processes* (process structure, operation types) -> *configuring the resource* (structure of system, equipment and staff types)". An abstract product – process – resource model is based on the concepts of ontology – oriented constraint networks. The dynamic constraints network is a model to solve constraint satisfaction problem represented by ontology of entities. Based on concepts of the dynamic constraints network, the taxonomy of product-process-resource model is elaborated in figure 9-6.

The taxonomy of product-process-resource model represents the tree of taxa, set of characteristics of supply chain, combined by their internal homology. Processes of supply chain are described in taxonomic representation as flows and transformation synthesis, which are transformation of flows. Flows of supply chain are financial, information, material, and product. Depending on the selection of elements from product-process-resource model taxonomy, the supply chain process model construction will change.

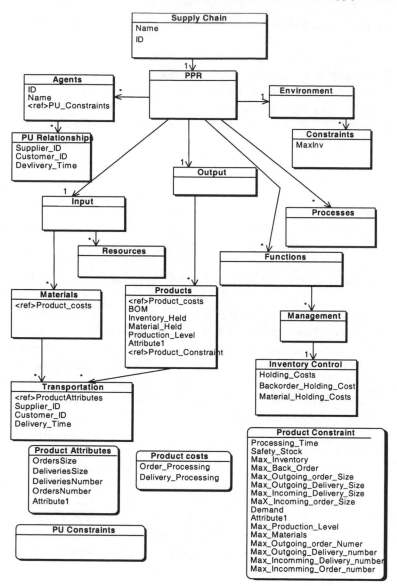

Figure 9-5. Information model: Ontology

4.2.4 Supply chain process model construction

The supply chain process model entity can be considered as a system with corresponding parameters related to input, output, function, rules, agents, processes, and environment. It has a production unit entity. Product entity is an element of supply chain that serves as an output of production unit entity. Entities of supply chain process model contain two types of attributes, which are properties and characteristics. Properties are attributes of entity describing input, output, function, and processes parameters of system and providing information about entity as a separate unit. Characteristics are attributes of entity describing rules, agents, and environment, and providing information about role of entity in the supply chain. Relationships between entities of supply chain process model are met with some restrictions. For example, two types of relationships have been implemented in this research, viz., those between product and production unit, and production units themselves (Curtis et al. 1992, Johannesson and Perjons 2001, Lambert and Cooper 2000, Martin and Cheung 2000, and Landauer 2000).

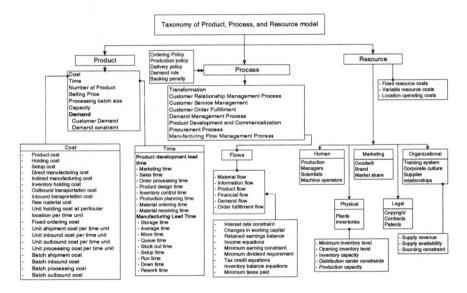

Figure 9-6. Product-Process-Resources Model Taxonomy

4.2.5 Ontology Development

We advocate the use of Ontology as a means of bridging domain analysis (taxonomy) and application system construction (or decision modeling system). Our approach is to consider ontologies as the basis for specifying models in a specific problem domain (forecasting management, inventory control, production scheduling, etc.). The scope of ontologies is restricted to a particular problem domain, which permits assumptions to be made with regard to system architecture related to the problem-solving environment. On this basis, concepts in the ontology can be explicitly linked to software component capabilities, enabling the ontologies to serve both as mechanism for indexing relevant software components and as specification of overall configuration requirements.

For example, consider an inventory level optimization problem for which we need to create an information model. But before building a model, we have to collect characteristics, which describe this problem. The focus of our study is how to make those characteristics reusable and applicable for solving other problems that arise in the supply chain environment. The idea of creating ontology is to create a repository of characteristics grouped in object-oriented hierarchy. Ontology $A_i = (V_i, D_i, C_i)$ is a static constraints network, which contains three parts: variables V_i taken from particular domains D_i, and constraints C_i for these domains. Utilizing these characteristics, we can start building information meta-model by:

- Starting with system taxonomy structure,
- Finding classes, where above mentioned characteristics exist and selecting them for the domain problem,
- Building product-process-resource information model, utilizing UML/XML diagram with classes taken from taxonomy, and
- Giving initial values to characteristics.

As a result, an information presentation format is developed, which can be used by the decision model. There are two main reasons for utilizing these steps, instead of using the plain list of characteristics. First, these steps offer a standardized format needed by computational tools, which is common for every problem-solving environment. Second, these steps allow reusability of the same structure, whereby any other module can use results from this problem-solving module. Ontology information structure meta-model for Inventory Control is depicted in figure 9-5.

Supply Chain Management concept is an approach to industrial network enterprise creation and reuse that considers enterprises as assemblies of reusable units defined on shared "product–process-resources" domain knowledge model. Each object of above model represents knowledge about an agent charged with delivering a specialized technology. For example, the supply chain agent is composed of one or more enterprise agents. Enterprise agent is composed of one each of inventory manager, capacity manager, and production manager agents. Similarly, Inventory Management agent is composed of one each of forecast management, inventory control, and raw material management agents. This relationship between agents signifies coordination of strategies, policies, goals, and objectives among them for problem-solving in specific domain.

Figure 9-7, depicts an object-oriented domain problem-solving / service ontology model for the inventory management agent. Its main components are inventory control, forecast management, and raw materials management agents, each of which carries specialized knowledge about these expertise areas / topics. For a supply chain, the object-oriented domain descriptions are as follows:

- Object supply chain describes the specific domain product supply chain agent.
- Object enterprise describes various member agents for this particular product supply chain, i.e., retailer, assembler, component manufacturer, and end-product manufacturer, etc.
- Objects inventory, capacity, and production describe agents with specialized knowledge in these fields.
- Objects FM, IC, and RMM describe agents with domain knowledge in the areas of forecasting management, inventory control, and raw materials inventory management, specific to inventory management.

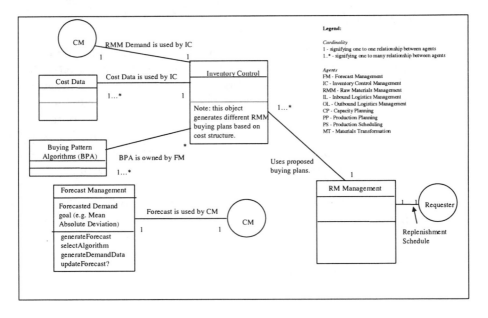

Figure 9-7. Inventory Management Domain Object Problem-solving / Service Ontology Model

Each agent is characterized in terms of services. A description of services for forecast management agent is provided in Table 9-1 with corresponding ontology class. The service description is XML codified. Agents register their services with the DF (domain facilitator), depicted in figure 9-7. All offered services are invoked by sending a message to the corresponding agent. Implementation of the communication language permits defining multiple ontologies in the message parameter section.

In this service "forecast", both Service Ontologies and Ontology of Administration and Management are invoked.

Table 9-1. Service "Forecast" description for invoking Ontologies

Agent: sub-agent Forecast:
:*service-name* generateForecast
:*service-description* Utilizing demand data generate a forecast time series.
:*service-pre-defined constants* time (period) = 52
:*service-type* Data manipulation
:*service-ontology* ForeacastManagement
:*service-fixed-properties* forecast_error_max
:*service-negotiable-properties* forecast_error_min, price
:*service-communication-properties* FIPA-Iterated-CNP, FIPA-Auction

5. CONCLUSIONS

Knowledge representation by ontology driven dynamic constraint networks (such as a supply chain) makes it feasible to provide powerful tools for knowledge management for mass customizable products. The knowledge management environment provides means for supply chain ontology-oriented collaborative engineering. Ontology-based architecture supports co-operation among agents, thereby reducing time and increasing the quality of supply chain configuration process. Ontology-based knowledge management technology is an innovative technology in the supply chain domain. Using this technology enables improved decisions on supply chain configurations from supply chain unit templates, under constraints networks with reduced variance. Implementation of this technology will enable realise increased quality, reduced cost, reduced errors, decreased personnel requirements, and better supply chain configuration solutions, which are an essential ingredient for the mass customisable production approach.

Future research on this topic will certainly lead to development of generic supply chain network configurations for a mass customizable production environment, based on, (a) divergent problem solving strategies, (b) functional or operational policies of Members and / or Group, and (c) levels of co-operation among members of the supply chain.

REFERENCES

1.   Bartalanffy, L. V. *Perspectives on General system Theory.* New York: George Braziller, 1975.

2. Curtis, B., Kellner M. I., and Over J. Process modeling. Communications of the ACM 1992, 35/9: 75-90.
3. Chandra, C., Enterprise architectural framework for supply-chain integration. Proceedings: 6th Annual Industrial Engineering Research Conference; 1997 May 17-18: pp.873-878, Miami Beach, Florida, Institute of Industrial Engineers, Norcross, Georgia.
4. Chandra, C., Smirnov, A. V., and Chilov, N., Business Process Reengineering of Supply Chain Networks through Simulation Modeling & Analysis. Proceedings: Second International Conference on Simulation, Gaming, Training and Business Process Reengineering in Operations; 2000 September 8 – 9, Riga, Latvia.
5. Da Silveira, G., Borenstein, D., Fogliatto, F. Mass customization: Literature review and research directions. International Journal of Production Economics 2001; 72/1: 1-13.
6. Foundation for Intelligent Physical Agents. *FIPA 98 Specification*, 1998. Part 12 – Ontology Service. http://www.fipa.org.
7. Fikes, R. and Farquhar A., Distributed repositories of highly expressive reusable ontologies. IEEE Intelligent Systems & Their Applications 1999; 14/2, Mar-Apr: 73-79.
8. Fischer, K., Müller, J. P., Heimig, H., and Scheer, A.-W., Intelligent Agents in Virtual Enterprises. Proceedings: First International Conference and Exhibition on the Practical Application of Intelligent Agents and Multi-Agent Technology; 1996: 205-223; The Westminister Central Hall, London, UK.
9. Fox M. and Gruninger M., *Ontologies for Enterprise Integration*. Department of Industrial Engineering, University of Ontario, 1999.
10. Gasser, L. Social conceptions of knowledge and action: DAI foundations and open systems semantics. Artificial Intelligence 1991; 47: 107-138.
11. Gruber, T. Toward principles for the Design of Ontologies Used for Knowledge Sharing. International Journal of Human and Computer Studies 1995; 43/5(6): 907—928.
12. Gruninger, M. Integrated Ontologies for Enterprise Modelling. Enterprise Engineering and Integration. Building International Consensus; 1997, 368—377; K.Kosanke and J.Nell, Springer.
13. Hirsch, B., Information System Concept for the Management of Distributed Production. Computers in Industry 1995; 26: 229 – 241.
14. ISO TC 184/SC 5/WG 1, 1997, "Requirements for enterprise reference architectures and methodologies," http://www.mel.nist.gov/sc5wg1/gera-std/ger-anxs.html.
15. Jennings, N. R., On agent-based software engineering. Artificial Intelligence 2000; 117/2: 277-296.
16. Jennings, R., Cooperation in Industrial Multi-agent Systems. World Scientific Series in Computer Science 1994; 43, World Scientific Publishing Co. Inc.
17. Jennings, N., Faratin, P., Johnson, M., Brien, P., Wiegand, M. Using Intelligent Agents to Manage Business Processes. Proceedings of the International Conference "The Practical Application of Intelligents and Multi-Agent Technology; 1996, 345—360; London.
18. Johannesson, P. and Perjons, E. Design principles for process modeling in enterprise application integration. Information Systems 2001; 26: 165-184.
19. Lambert, D. M. Cooper, M. C. Issues in supply chain management. Industrial Marketing Management 2000; 29: 65-83.
20. Landauer, C. Process modeling in conceptual categories. Proceedings: 33rd Hawaii International Conference on System Sciences; 2000.
21. Lee, H. L., and Billington, C. Material management in decentralized supply chains. Operations Research 1993; 41/5: 835-847.
22. Lesser, V. R. Cooperative Multiagent systems: A personal View of the State of the Art. IEEE Transactions on Knowledge and Data Engineering 1999; 11/1: 133-142.

23. Manufacturing Enterprise Integration Program (MEIP). *National Institute of Standards and Technology (NIST)*. Gaithersburg, Maryland. http://www.atp.nist.gov/, 1999.
24. MESA International White Paper # 6. *MES Explained: A High Level Vision*. http://www.mesa.org, 1998.
25. Martin, I. and Cheung, Y. SAP and business process re-engineering. Business Process Management Journal 2000; 6/2: 113-121.
26. McCarthy, I. Manufacturing classification. Integrated Manufacturing Systems 1995; 6/6: 37-48.
27. McCarthy I. Ridgway K. Cladistics: a Taxonomy for manufacturing organizations. Integrated Manufacturing Systems 2000; 11/1, 16-29: 16-29.
28. McKelvey, B. *Organizational Systematics Taxonomy, Evaluation, classification*. Berkeley: University of California Press, 1982.
29. National Industrial Information Infrastructure Protocol (NIIIP). www.niiip.org, 1994.
30. Rubenstein-Montano B., Leibovitz J.,Buchwalter D., McCaw B., Newman K., Rebeck K. A System thinking framework for Knowledge management. Decision Support Systems 2001; 31: 5-16.
31. Sandholm, T. Agents in Electronic Commerce: Component Technologies for Automated Negotiation and Coalition Formation. Proceedings: International Conference on Multi Agent Systems; 1998 10-11, Paris, France.
32. Smirnov, A. DESO: GDSS for Virtual Enterprise Configuration Management. Proceedings: 5th International Conference on Concurrent Enterprising ICE'99; 1999; The Hague, The Netherlands.
33. Smirnov, A. V. Conceptual Design for Manufacture in Concurrent Engineering. Proceedings: Conference on Concurrent Engineering: Research and Applications; 1994, 461-466; Pittsburgh, Pennsylvania.
34. Smirnov, A. V. Chandra, C. Ontology-based Knowledge Management for Co-operative Supply Chain Configuration. Proceedings: 2000 AAAI Spring Symposium "Bringing knowledge to business Processes"; 2000; March 20-22: 85-92; Stanford, California, AAAI Press.
35. Sousa, P., Heikkila, T., Kollingbaum, M., and Valckenaers, P. Aspects of co-operation in Distributed Manufacturing Systems. Proceedings: Second International Workshop on Intelligent Manufacturing Systems; 1999, September: 685-717; Leuven, Belgium.
36. Uschold, M. Gruninger M. ONTOLOGIES: Principles, Methods and Applications. Knowledge Engineering Review 1996; 11/2.
37. Van Hoek, R. I. The rediscovery of postponement a literature review and directions for research. Journal of Operations Management 2001; 19: 161-184.
38. Vasconcelos J., Kimble C., Gouveria, R. F. *A Design for a group Memory System using Ontologies*. UKAIS, University of Wales Institute, Cardiff, 2000.
39. Work Flow Management (WFM) (1996). www.wfmc.org,
40. Wooldridge, M. Jennings, N. R. *Intelligent Agents - Theories, Architectures, and Languages*. Lecture Notes in Artificial Intelligence, Springer-Verlag, 1995.

SECTION 4:

FUTURE RESEARCH AGENDA

CHAPTER 10

FUTURE DIRECTION ON MASS CUSTOMIZATION RESEARCH AND PRACTICES: A RESEARCH AGENDA

Janet Efstathiou, Ting Zhang

University of Oxford

Abstract: This chapter identifies and reviews the issues for the mass-customizing manufacturing organization, in terms of quality and flexibility of processes, design of products, customization within the supply chain, performance measures, inventory strategies and ICT. We present a discussion on the competencies that an organization is likely to need for success with mass customization. We conclude with some research questions to test scientifically theoretical models of mass customization, and a summary of the elements of mass customization strategies.

Keywords: Competencies, complexity, supply chain.

1. INTRODUCTION

Mass customization has been advocated for some time as the next step forward in the evolution of manufacturing systems. Much of the advocacy for this seems to have come from the marketing and business school perspectives (e.g. Duray 1997, Pine 1993). However, Mass Customization is perhaps an evolutionary stage in the development of manufacturing systems and philosophies over the past fifty years. We had the Total Quality movement, and the interest in Just in Time manufacturing which led eventually to lean and agile manufacturing. All these movements with their emphases on quality, efficiency, minimizing waste and responsiveness to the customer, were necessary enablers and pre-cursors for the ideas and practices of Mass Customization. However, a significant enabler of Mass Customization is the Internet, providing customers with the power to order customized products, and the supply chain with the ability to communicate rapidly.

While we can point out the contributions from past theory and experience, we need to look ahead to uncover the research agenda for the coming decade so that industry can be poised to take advantage of the new opportunities that will become available. The focus of this chapter will be on competencies that manufacturing and Mass Customization organizations will need to understand and adopt in the next ten years.

The chapter is organized as follows. Section 2 deals with issues that manufacturing systems will face as they strive to adopt Mass Customization. Section 3 covers issues from the organization's point of view. Section 4 suggests research questions that should be considered in the future in order to understand Mass Customization rigorously. Section 5 reviews components of an organization's Mass Customization strategy and looks briefly at some of these issues over the horizon.

2. ISSUES FOR MANUFACTURING

We highlight in this section the concerns that will be of particular importance to production management. These include the broad areas of product design and innovation, process design and planning, and production scheduling, monitoring and control. Although these functions have all seen innovations from Information and Communications Technology (ICT), many organizations in the extended supply network still rely on people and their flexible skills to cope with the dynamic environment that will characterize Mass Customization.

In this section, we will consider the following topics:
- Quality and flexibility of processes

- Design of products
- Customization within the supply chain
- Performance measures
- Inventory strategies
- ICT to support Mass Customization ordering, product configuration, production management and the supply chain.

Before proceeding to discuss specific issues, we shall consider some of the current issues facing the development of Mass Customization. They may be summarized as the nature of customization and the cost of production. According to Lampel and Mintzberg (1996), a customized product is designed specifically to meet the needs of a particular customer, although variety is needed to provide choices. At the same time, an ability to specify the customer's need is also important (Duray et al 2000). The second issue refers to the methods of realizing mass customization at or near mass production costs (Tseng and Jiao 1996). Pine (1993), Sako and Murray (1999) and Duray *et al.* (2000) suggest the practice of modularity, which provides the means of producing components in volume as standard modules with product distinctiveness achieved through combination or modification of modules. Therefore, modules that will be used for the customized product can be manufactured using mass production techniques and achieve near mass production costs.

Although the concept of Mass Customization has been discussed in the literature for more than a decade, it has only been recently that the necessary improvements in manufacturing systems and product design have enabled the concept to become reality. To achieve this, manufacturing had to reduce the trade-off between variety and productivity (Ahlström and Westbrook 1999, Kotha 1995, Pine 1993, Victor and Boynton 1998). Furthermore, the widespread take-up of new ICT and the Internet was another important enabler of Mass Customization. Indeed, some authors (e.g. Piller and Moeslein 2002) argue that Mass Customization is little more than an application of e-commerce. This implies that the mass-customizing enterprise must have the capability and resources to manage effectively the high information input from customers, and the onward dissemination of that information to suppliers.

It almost goes without saying that the mass-customizing organization needs to be a high quality manufacturer that can advertise with confidence the quality and reliability of delivery of its product. Customers who are taking the time and commitment to request a customized product are unlikely to give the supplier a second chance if the delivered product is not up to their expectations. Also, the manufacturer that does not already have these competencies is unlikely to be able to aspire to become a mass customizer. In order to assist the customer to make the right product choices, mass-customizing manufacturers will need to have some kind of product configurator that will allow the customer to explore the options and check that their customized product choice is exactly what they intend, and that it can be manufactured.

The mass-customizing manufacturer will have to make a decision on where in the manufacturing process to allow the customer to become involved. A number of authors have considered the customer involvement point as a key element of MC and have used it as a means of classifying different approaches. Gilmore and Pine (1997) classify MC based on the degree of change in products and degree of change in the representation to the customer. They define the types of mass customization as "faces" and categorize MC according to four faces: cosmetic, collaborative, adaptive and transparent. Alford et al (2000) focus on the manufacturing process in the automotive industry, and note the point at which variety is introduced into the product, called the customer involvement point or decoupling point. Spring and Dalrymple (2000) further develops Alford's classification and emphasizes the continuum from mass production to pure customization. They identify five decoupling points and map these onto five strategies for Mass Customization. The decoupling points normally take the form of inventory and mark the point between a mass production push line and the mass customization pull line. Their customization strategies may be summarized as follows:

- Post-Distribution: no customization, or customization carried out after distribution,
- Distribution-to-Order: the customer has package options at the delivery point.
- Assemble-to-Order: fabrication is carried out prior to order and inventory of raw materials, modules and components is held.
- Fabricate-to-order: only raw material inventory is held.
- Engineer-to-order: no inventory is held, so the customer is involved at every stage.

This classification of MC strategies will be used later in this chapter, in Sections 2.5 and 3.3.

2.1   Quality and Flexibility of Manufacturing Processes

In a mass-customized environment, the quality of the manufacturing process is particularly important. The performance, monitoring and control of a mass-customized production line is much more sensitive to the presence of defects than was the case with a mass-production facility.

During the manufacturing process, products flowing along the production line may be unique and configured to a particular customer's requirements and specification. The customizing stages of the manufacturing process may be done in batches of one. If a defect is introduced onto that product, it is not possible to replace it with another item from the same batch or the next item on the line. If an item has to be removed from the line for rework or repair to remove a defect, then the flow of production will be interrupted twice – once, leaving a space in production where the defective item was removed and, secondly, pushing items back in production in order to create a space for the re-introduction of the repaired item to the production sequence. Defective items in a Mass Customization manufacturing environment have the potential to severely disrupt the delivery of products on time to customers' requirements. Hence, the occurrence of defective items during production should be avoided because of the disruption it causes to the planned delivery of products to the customer, the possible delays caused to other customers when the defective item is re-introduced and the problems to the manufacturer of tracking the position of every item in the planned sequence.

Another problem with the manufacture of unique products is the difficulty of diagnosing and removing intermittently occurring faults. For example, a computer manufacturer produces computers, each uniquely configured to a customer's requirements. The lifetime of computer components is very short, in some cases as short as six weeks. Hence, a unique computer can be assembled, tested and delivered to customers, only to develop a fault in the field. Tracking down and identifying the fault to the co-occurrence of two components that perhaps had only been assembled together in a small number of computers, due to their short lifecycles, is an exceedingly difficult and complex task, requiring very well developed computer support. In this case, the quality issues are to do with the specification of the complex components and modules that make up the finished product. The testing and burn-in procedures must also be thorough, but fast enough not to compromise the user's desire for a reliable and customized product with a lead-time and price that do not differ very much from the mass-produced product.

In order to cope with the high variety of product that is passing through the production process, the manufacturing processes themselves need to be flexible and reliable. There is evidence from studies of the impact of variety that high levels of product variety can have an impact on the quality and reliability of the manufacturing process, leading to poor quality of product and higher probability of breakdown of machines and resources. The Total Quality movement has long advocated the reduction of variety as a means of improving quality, so to strive for more variety would seem like a rod for one's own back. However, we must learn and apply the lessons from Total Quality as we move gradually towards full Mass Customization, such as scheduling and carrying out proper maintenance, investing in the appropriate tools and equipment, and ensuring the training and motivation of the workforce.

2.2   Design of Products and Product Mix

It is widely agreed that Mass Customization can be achieved by using modular components that can be assembled to the customer's specifications as late in the manufacturing process as possible (Sako and Murray 1999). For many products, this will not be true (e.g. some furniture, some cars etc.), but for other products the late configuration model is a viable way to achieve Mass Customization. The would-be customizer must be wary though.

An important issue for the product designer and manufacturers to consider is the desirability in the market of the options that are being offered. The perceived value to the customer of particular combinations (or the possibility of ordering particular combinations) must be weighed up against the practical consequences of enabling the manufacture of those components on the shop floor.

Mass Customization offers manufacturers a rapid view of what the market seeks and by making and delivering the product quickly, the manufacturer should be better equipped to anticipate fashions and shifts in demand. However, to exploit this information, the manufacturer has to be prepared to adapt the product range quickly, introducing new products and rigorously pruning the less desirable in order to match effectively the demand that emerges from the marketplace. We shall return to the consequences for inventory strategy in section 2.6.

2.3   Customization within the supply chain

While the manufacturer wishes to customize as late as possible, it is desirable to add the customized value within one's own manufacturing facility, and use standard components supplied by outside suppliers. This is the case with the computer supplier, who produced individually configured computers or computer networks while using standard components supplied in bulk. However, for some manufacturers and products, the nature of the product and the degree of customization may imply that some degree of customization has to be done by suppliers.

This has important implications for the manufacturer, especially with respect to lead-time, i.e. the time from the customer placing an order for the product to delivery of the goods. If customization has to be carried out within the supply chain, this will mean that the customer's order has to be prepared and submitted to the supplier, who must then customize or configure the component and deliver it to the assembler. Clearly, this will add time to the manufacturing process and will only be acceptable for some industries that tend to have long customer lead-times. Thus, it would be (and is) workable in the automobile industry, but would require special configuration of the supply chain in the fast-moving consumer electronics sector. Typically, one would expect to see close geographical co-location and integration of the ICT systems for such a supply chain that involves supplier-customized modules to be effective.

A number of different supply chain structures are possible for products involving supplier-customized modules. A few are depicted in Figure 10-1. The most familiar structure is In-house Customization, where the assembler carries out the customizing activity, using standard components supplied by various non-customizing suppliers. The point at which customization is introduced in the assembly process is an important factor in the ability of the assembler to deliver the optimal amount of variety within an acceptable consumer lead-time. A second strategy is Early, Minor Supplier-Customization, where most of the components are standard, but a customizing supplier supplies some components that are used early in assembly. This strategy would imply a longer customer lead-time, because of the impossibility of proceeding with any in-house assembly until the supplier delivers the customized item. It also implies that the ability of the assembler to proceed with the manufacture and delivery of a customized product depends strongly on the quality of the products and reliability of delivery provided by the customizing supplier. A preferred strategy would be Late, Minor Supplier-Customization. In this case, assembly can proceed while the supplier is preparing the customized component. This would have a better consequence for customer lead-time and leave the supplier less vulnerable to delays in goods arriving from the supplier.

Another difficult strategy is Late, Major Supplier-Customization. This involves the supplier preparing a customized component that forms a major part of the product, but that takes nearly as long to prepare the in-house assembly done by the assembler, as shown by the length of the arrows in the diagram. This kind of strategy would be best suited to a close, co-operative partnership arrangement, since both the assembler and supplier are adding large amounts of value and both want to ensure that the customer gets the product that they expected.

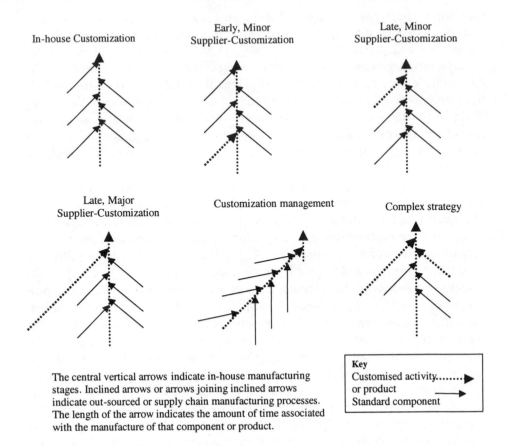

The central vertical arrows indicate in-house manufacturing stages. Inclined arrows or arrows joining inclined arrows indicate out-sourced or supply chain manufacturing processes. The length of the arrow indicates the amount of time associated with the manufacture of that component or product.

Figure 10-1. Proposed supply chain strategies for customization

An alternative strategy is to go for Customization Management. In this case, the "assembler" does very little assembly at all, but buys in a near-complete product from the customizing-supplier who carried out the task of managing the component-suppliers and organizing the supply chain. In this case, the "assembler's" role has become that of the marketer and customer interface, transmitting orders onwards to the customizing-managing-supplier and managing some parts of the order fulfillment process

The final strategy depicted in Figure 10-1 is a complex strategy that involves a mixture of standard and supplier-customized components. Many other complex strategies are possible, involving elements from all five other supply chain strategies.

2.4 Performance measures

Many performance measures have been suggested and recommended for use in manufacturing systems. We suggest that the key performance measures for the consumer of a mass customized product are:

- Price to consumer
- Consumer lead-time.

It is, of course, possible to elaborate both these measures further, to include expected lead-time, maximum lead-time etc. One could also extend the measures for customer satisfaction, such as probability of satisfying from late customized inventory or from work-in-process. However, these are the key performance measures that distinguish a mass-customizing manufacturer from a mass-producing manufacturer.

Note that we have not included the number of options offered to the consumer as one of the performance measures. In a sense, this is an input parameter. By setting the range of options that the product will offer, the manufacturer must design the assembly process and the supply chain in order to deliver the acceptable price and lead-time. Each consumer is only interested in satisfying their own needs, so there is only one option that they are each interested in. The fact that the assembler may be able to offer thirty, thirty thousand or a very large number of different configurations (as is the case in the automobile industry (Holweg and Pil 2001)) is a means of advertising their competence at managing a customizing manufacturing process and supply chain.

Other important performance measures for a mass-customizing organization will include the quality, lead-time and responsiveness of their suppliers. These factors will be particularly important if the suppliers undertake part of customizing work, as seen in Section 2.3.

2.5 Inventory Strategy

Lean Manufacturing and Just in Time manufacturing have emphasized the need to reduce the amount of material held in inventory. However, in order to deliver quickly to customers, it is necessary to hold some stock in order to buffer out the variation in consumer demand. For some customizing manufacturers, especially those at the Fabricate-to-Order and Design-to-Order end of the spectrum, there is scope only to hold raw material stock and it is impossible to hold any stock that has had value added. For customizing manufacturers who wish to Assemble-to-Order, then it may be possible to hold stock that has had some value added, with only some late configuration required.

The inventory strategies that are available to the customizing manufacturer have to take into account the structure of the product and the desired lead-time. This is illustrated in Figure 10-2. It is likely that the Assemble-to-Order customizer will have to hold some stock in order to respond rapidly to variable demand. We assume, for now, that the stock acts as the decoupling point between the pull process from the consumer and the push process from the low variety manufacturer. We assume that the manufacturing process involves a number of stages. The closer these are to the consumer, the more likely they are to differentiate the product into an ever more narrowly specified range of possible product variants.

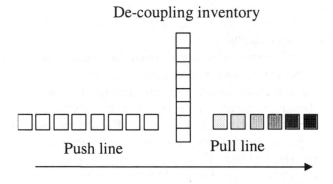

Figure 10-2. Schematic of the Mass Customization manufacturing process

[Undifferentiated products are made in the Push line, which are then stored in inventory. When a customer order is received, products are taken from inventory and customized in the Pull line.]

In terms of lead-time, the best strategy is to locate the stock as close to the consumer as possible. The lead-time will involve only locating the exact product the customer requests from stock, and dispatching the order to the customer. From the manufacturer's point of view, this is a less desirable option, since it involves holding in stock examples of every item from the catalogue of options. This can be a problem where items go out of fashion or deteriorate in storage. Holding stock as complete products implies a large amount of capital is tied up. This option could only be possible for small, low variety, fast-moving items.

As the stock holding point is moved back along the manufacturing pipeline, the lead-time as seen by the consumer will increase, since there are now some stages of the manufacturing process that must be completed, in addition to the process of retrieving the item from stock. This process becomes more desirable from the manufacturer's point of view, since less value has been added to the items that are in stock. Another important point is the fact that the stock now contains items that are less differentiated, so that there are fewer variants that have to be managed. When the stock is held at the distribution point, there are, say, n variants, with stock held further up the manufacturing pipeline, there may be only m variants, where $m < n$. When the customer requests a product from the n variants in the catalogue, the manufacturer need only select from the m variants in stock and customize that item to turn it into the customized product that the consumer requires.

Recent work (Zhang 2003), has developed the model to calculate the *complexity* of the manufacturing process, where complexity is defined according to information theory as the expected amount of information needed to know the state of the system (Frizelle and Suhov 2001). Any mass-customizing organization will have to deal with a large amount of information from customers and suppliers, and so it will be desirable to reduce the amount of information generated internally. This will require the minimum possible amount of information consistent with achieving the goals of Mass Customization. Zhang's results (Zhang 2003) show that the two factors most likely to increase manufacturing complexity in a mass-customizing manufacturer are:

1. The number of options available to the customer
2. The number of items held in inventory at the decoupling point.

Other factors have been included in the model, such as stock holding strategy, number of manufacturing stages after the customer involvement point and the number of semi-finished variants held at the customer involvement point. Their relative effect has been shown to depend on the actual circumstances of the manufacturer. Further details may be found in Zhang (2003).

Altogether, the choice of product design, manufacturing and stock holding strategy, therefore, involves a balance between
- the number of semi-customized variants that are to be held, m,
- the position of stock in the manufacturing pipeline,
- the added-value in the pipeline at that point,
- the amount of each variant to be held, and
- the desired level of customer satisfaction.

2.6 ICT to support Mass Customization ordering, product configuration, production management and the supply chain

Clearly, we have yet to appreciate and apply the full possibilities of the Internet on manufacturing and Mass Customization in particular. Just as TQM, JIT, lean and agile manufacturing were essential ingredients for Mass Customization, we cannot now conceive of MC without the Internet to enable the fast exchange of data with the customer and subsequently along the manufacturing supply chain. In this subsection, we shall outline briefly some of the issues facing manufacturing systems in the coming decade.

Customers are already familiar with ordering some products over the Internet, such as books and groceries. A basket of groceries already has many of the characteristics of a Mass Customized product. It is uniquely configured for each customer, the price tag is about the same as for the same basket bought in the supermarket, the lead-time is similar (depending on the frequency with which one is accustomed to shopping) and the quality of the product is as good as from the supermarket. The customizing-manufacturer is here replaced by the retailer, who can use their large stock in the supermarket to avoid the need to tackle the supply chain any differently. The retailer can also take advantage of the deeper knowledge available on the consumer's habits to configure product families in such a way as to make the interaction with the consumer valuable and effective for both parties.

The mass-customizing manufacturer will also have to learn how to approach the mass market in a similarly attractive way. This will mean identifying and presenting product options that are attractive to the consumer while remaining feasible in the product structure and not too costly to store as a semi-customized product. The manufacturer and consumer will also have to understand the contract of ordering a mass-customized product over the Internet, where it is not always going to be possible to try the product, or an equivalent, for look and feel prior to taking delivery.

However, the closer relationship with the consumer will create the potential for knowing the consumer better and configuring products that are more likely to meet customer demands. As we saw with the example of the computer manufacturer, we need to keep a careful record of all known combinations of components that are likely to lead to an instance of a product that is likely to run into problems once it is operational. Hence, product configuration is an important task, both for ensuring the quality and reliability of the product and enabling the consumer to configure a valid and functional product. We see a similar trend in the development of customer loyalty cards for many of the large supermarket and clothing retail chains, as well as the customization of consumers' access to the websites of the on-line supermarkets.

Once the consumer has placed the order, associated orders could also be transmitted to stock and suppliers so that the manufacturing process can commence immediately, enabling the rapid and reliable delivery that we seek to the consumer. This high degree of automation can best be achieved with simple products with a limited range of options, at this time, but it is likely that further, higher levels of customization are likely in the future. In that case, the choice of suppliers will be very important, to ensure that they can deliver the quality of product, speed and frequency of delivery that are necessary for an acceptable consumer lead-time.

Armed with the possibility of fully automating the ordering, assembly and delivery process, we can tackle automating production management. By this, we mean more than the scheduling that is implied by the automated ordering process, but also the production planning, i.e. deciding what product mix and volume to accommodate.

In this section, we have considered some of the issues and competencies on the shop floor and supply chain that will have to be addressed by new entrants into mass customization arena. The competencies may be summarized as:

- High quality and flexible order fulfillment and manufacturing processes
- Ability to design Mass Customizable products
- Ability to support and manage the customizing supply chain
- Ability to co-ordinate with supply chain partners on identifying and applying appropriate performance measures
- Ability to manage inventory flexibly and accurately

In the next section, we shall look at some of the wider implications for the organization as a whole.

3. ISSUES FOR THE ORGANIZATION

In this section, we shall look at some of the consequences for an organization as it seeks to move from mass production to Mass Customization. Amongst these are:

- adoption of new performance measures
- dealing with value-adding customizing-suppliers
- identifying and developing the competencies for Mass Customization
- developing the business case for Mass Customization.

3.1   Adoption of New Performance Measures

There are two ways in which businesses will have to adapt to new performance measures. The first is that they will have to adapt to measures that emphasize consumer lead-time and cost. If the organization had hitherto sought to achieve any of the large lot sizes, high machine utilization or low stocks, they will likely have to think again. Short consumer lead-times will imply some spare production capacity, some stock and small lot sizes. This might require a cultural change for the organization.

The second way in which new performance measures may have to be adopted by the organization is in the situation where they are in a customizing supply chain, whether as a customizing-supplier or as a customer of a customizing-supplier. In this case, the supply chain will have to be tightly integrated in order to achieve the reliability of delivery and quality of product that are necessary. In order to achieve the tight integration, organizations will have to work to the same schedule along the supply chain. This may mean that a supplier may lose some autonomy in scheduling its production, since it must meet the schedule agreed with the customer. It would be most desirable if the consumer could set the schedule, but since this may not always be the case, it is likely that the bottleneck node will determine the ability of the whole supply chain to meet the desired consumer lead-times. Hence, schedules and associated performance measures will need to be agreed along the supply chain at the outset of an MC project if the required performance is to be met.

3.2   Dealing with Value-Adding Customizing-Suppliers

As we have seen in the subsection above, the organization will have to cope with the constraints of the customizing-suppliers, in terms of schedules and performance measures. The partnering relationship that is implied by these considerations may be a new experience for some manufacturers, and they may have to adjust to this unfamiliar situation.

We examined the structure of mass-customizing supply chains in section 2.3. The organization will have to think carefully about what structure of customizing supply chain it wants to create, and identify the risks and opportunities associated with each. They will need to consider whether their existing or intended product range is suitable for a supply chain that involves significant out-sourcing of value-adding activities, given the performance measures that will be required for Mass Customization. There will also be issues concerning the location of stock and at what point in the supply chain or in-house manufacturing process the customized or semi-customized items will be held and owned.

3.3   Identifying and developing the competencies for MC

We have suggested elsewhere (Mchunu et al 2002) that the competencies for a Mass-customizing organization may be summarized as follows:
- Product design
- Logistics and Information Management
- Distribution of Inventory
- Supply Chain Agility
- Flexibility

We also suggest that the required degree of competence in each of these different factors differs according to the MC strategy that the organization intends to adopt. In particular, we propose that the necessary degree of competence may be summarized as in Table 10-1. Each strategy will be discussed in detail below.

Table 10-1. Analysis of the competencies needed for different Mass Customization strategies.
[The level of competence is distinguished as High, Medium or Low to indicate the level that would be expected with respect to competing organizations in the same business sector.]

	Mass Customization Strategy				
Competency	*Post-Delivery*	*Distribute-to-Order*	*Assemble-to-Order*	*Fabricate-to-Order*	*Design-to-Order*
Design	high	low	medium	low	high
Flexibility	low	high	high	high	high
Supply Chain Agility	low	medium	high	medium	medium
Distribution of Inventory	low	high	high	high	medium
Logistics and Information Management	medium	medium	medium	medium	high

Post-Delivery   This manufacturing strategy implies that any customization that is carried out is done by the customer after delivery of the product. For this strategy to successfully achieve a perception by the customer of a customized product, the product must be extraordinarily well designed, with a very acute understanding of the customer's needs and perceptions. It must also be the case that the product is designed well enough for the customer to be able to configure it to their own needs with the minimum of effort. This strategy does not call for high levels of competency in the

Flexibility, Supply Chain Agility or Distribution of Inventory, beyond what is usually accepted in the industrial sector. However, Logistics and Information Management should be at the Medium level in order to be able to accept customer orders and react with the required product in the time required to give the customer a "customized" service.

Distribute-to-Order. This strategy implies having the customer involvement point very late in the manufacturing process so that products can be distributed to the customer, possibly from a value-adding retailer involving some low-level customization of the product. This strategy does not involve a particularly design competency, since existing product designs may be used or modified to suit this manufacturing strategy. However, the flexibility and responsiveness of the manufacturing process must be high in order to respond to fluctuating demands. The competence on management and distribution of inventory must also be high, in order to make sure that the amount of stock held around the network of customer distribution points is high enough to provide an acceptable level of customer satisfaction, but does not become tired or old. Since there is little customization carried out during manufacture, Supply Chain Agility does not need to be high, nor does Logistic and Information Management, since the perceived added value is in responding quickly from the late customer involvement point.

Assemble-to-Order. This manufacturing strategy involves several stages of assembly once the customer has placed their order. The key factors in being able to make this strategy deliver a rapid, customized service involve Flexibility, Supply Chain Agility and Distribution of Inventory. Since some of the components or modules that may need to be included in the product may not all be held on site, the supply chain must deliver products rapidly and reliably, the in-house inventory must be managed well, and the entire manufacturing and supply chain must be flexible. Product design is not a key factor here, however, the product must be designed for manufacture so that it can be assembled rapidly. Logistics and Information Management, while important, is not rated as one of the key enablers for this manufacturing strategy.

Fabricate-to-Order involves the customer very early in the manufacturing process, after the design stage but prior to the fabrication. Here, the key competencies are Flexibility and Distribution of Inventory. Since the product is being fabricated to order, e.g. a customized bicycle, it is very important that the customized components that are required early in the manufacturing process can be obtained quickly and reliably. If the fabrication is done in-house, then Flexibility is very important, but Supply Chain Agility only needs to be medium. However, this position may change if fabrication is out-sourced, when the MC organization would require a high level of supply chain agility. If this cannot be guaranteed, then Fabricate-to-Order is unlikely to be part of a successful MC strategy. In this example, Design competency is not seen as a key enabler, beyond the levels usually seen in the relevant sector.

Design-to-Order. In this strategy, Design competence is absolutely crucial to achieve a rapid response to the customer, and to enable rapid and reliable manufacture. The manufacturing process must also be highly flexible in order to produce the customer's desired product once the design has been agreed. Because of the many stages that could be involved in agreeing the design, ordering the supplies and managing the work in progress, Logistics and Information Management competence must also be high. Although there is no or very little inventory held, Distribution of Inventory is an important competency, since here it covers management of the work in progress. Supply Chain Agility also need only be medium, since orders to the supply chain can be integrated with the design process.

Table 10-1 suggests the competencies that are required for different MC manufacturing strategies. IT may be used by organizations that are considering MC as a way to compare their current competencies with those required for the MC manufacturing strategies they would like to adopt.

For example, this could help companies avoid making the mistake of thinking that because another firm in their manufacturing sector has adopted a particular MC manufacturing strategy, that strategy ought to be adopted by them as well. What this table shows is that the MC manufacturing strategy depends on the competencies of an organization, with respect to the competencies normally achieved in their sector.

Hence, Table 10-1 may be used in two ways. First, an organization could carefully scrutinize its competencies and find the MC manufacturing strategy that is closest to its current level of competence. This will highlight the areas that need to be built up in order to achieve that particular MC manufacturing strategy, and help an organization avoid making a costly mistake when it discovers there is more to implementing a strategy than they had realized. An organization may get involved with Mass Customization by identifying more than one strategy that are already close to their range of competences, and set up separate manufacturing resources to pursue each strategy, following the idea of the focused factory (Kotha 1995) which can be used to bootstrap competence in MC and manufacturing performance.

Second, an organization could decide that its market demands a mass customized product and selects the manufacturing strategy it considers to be appropriate. Again, it must to a rigorous scrutiny of its competencies, and use a gap analysis to identify those areas where its current level of competence is not sufficient to achieve the desired strategy. This will help the organization develop the business case for adopting Mass Customization, while providing a systematic way of deciding how to close the gaps and achieve the multiple objectives needed for Mass Customization.

3.4   Developing the Business Case for Mass Customization

Many other chapters in this book will also have made reference to the business case for Mass Customization. We summarize the points of the case as follows:

1. Is there a market demand for customized products in our business sector?
2. What Mass Customization strategy is appropriate to our product range and market?
3. What competencies will we need to invest in to achieve the required standard?
4. Is the demand sufficient to warrant the investment required?

The competencies are wide-ranging and may imply a substantial change in the way the business is run and in how it perceives itself. There will have to be structural changes in the way that orders are obtained from customers and dispatched to the manufacturing process. Inventory strategies may have to be modified in order to achieve the goals of price and lead-time that need to be met.

It may not be possible (or wise) to aim for full Mass Customization in one step, but rather to approach it as a phased problem, developing the competencies and measuring progress towards the necessary standards before launching a full MC strategy.

It is important to look at the competition very carefully and identify their competencies. It may be possible to predict which MC strategy they are likely to adopt, and possibly estimate their probability of success. Hence, one could choose the strategy that best matches the competency profile and likely gap in the market.

Mass customization is a challenging strategy, but one that cannot be ignored. For some business sectors, away from the commodity market, it can offer opportunities of adding value for customers, through a combination of products and services that could create new opportunities and benefits.

4   Research Agenda

In this section, we bring together the issues from the previous two sections to identify some research questions for the future.

Where in the manufacturing pipeline should inventory be stored to balance cost and expected customer lead-time?

As we have discussed in sections 2.3, 2.5 and 3.2, some inventory is likely to be needed to bring down expected consumer lead-time. The product may have to be designed or configured in such a way as to enable the customizing to be done as late in the manufacturing pipeline as possible. It may be the case that there ought to be more than one location in the manufacturing pipeline where stock should be stored. Ideas from multi-echelon inventory theory will have to be balanced with product design to enable late value-adding onto semi-customized products.

How can the products options be identified that offer the most perceived added value?

This is a problem for the marketing departments as they study the range of options that people prefer in related product areas. Expectations need to be created that are within the range of feasible and cost-effective products.

How can the shift in observed customer purchasing decisions be quickly fed back to the push manufacturing pipeline in order to ensure stock and product design tracks changing fashions as accurately and quickly as possible?

One of the possible advantages of Mass Customization is the ability to obtain quick snapshots of customer demand. This valuable information needs to be incorporated into the manufacturing process, prior to inventory holding, and used to match demand and push manufacture as well as possible.

How can the complexity of Mass Customized manufacture be minimized and managed effectively?

The further up the manufacturing process that products are tagged with the identity of the consumer, the greater will be the complexity of the manufacturing process. In designing the ICT system to support Mass Customization, attention must be paid to how much information needs to accompany each item during its process of manufacture. Mass production is straightforward, because work in progress may only be identified as a batch that is to be sent to one customer. However, a mass-customized item needs to have the customer identity and the choice of options associated with it. The way to minimize manufacturing complexity is to leave that point of unique registration with an individual customer as late in the manufacturing process as possible. However, the ICT systems will have to be designed in order to achieve the most effective and manageable system that meets the manufacturer's demand for reasonable amounts of information at the appropriate precision and cost.

What are the factory layouts and IT configurations that best enable effective monitoring of the system state and the flexible manufacturing?

Many SMEs are capable of participating in Mass Customization supply networks, but they may not be able or willing to invest in ICT systems that are superfluous to their needs and more expensive than they can reasonably justify. There is a need to understand the precision with which the state of the manufacturing process needs to be monitored in order to ensure that the ICT systems are appropriate for all the supply network members as well as the task each member must perform.

What supply chain strategy is best suited to particular industries and product structures?

The policies for late customization, ICT management, consumer price and lead-time will differ from industry to industry, depending on the manufacturing process and costs, consumer preferences etc. While a theoretical model of the Mass Customization manufacturing process may be built up, this will have to be reinforced with carefully chosen case studies of mass-customizing organizations and their supply chains.

Case studies of Mass Customization

Although a few notable case studies of Mass-customizing organizations have already been carried out, it will be necessary to identify and conduct some more. These will be able to investigate both the manufacturing competencies and the ability of the organization's structure to adapt to a Mass Customization style of manufacturing. It is likely that global differences in customer preferences will have effects on the perceived market demand for customization, so case studies from different regions will be needed. It will also be necessary to populate the spectrum of mass customization manufacturing strategies, from Design-to-Order right through to Distribute-to-Order, combined with inventory strategies (Section 2.5) and supply chain strategies (Section 2.3) with cases of attempts to achieve Mass Customization. Indeed, research projects that involve consortia of universities and industries working together to share the findings would be a beneficial and instructive way to learn the lessons of Mass Customization while disseminating them as widely as possible.

How to test the competency model?

The competency model of Mass Customization speculates on what are the competencies that are likely to be needed by organizations in different regions of the mass-customizing spectrum. This can best be tested by comparing the competencies of organizations and their predicted success, as they embark on Mass Customization strategies, with the outcome. As with the previous research question, this can best be tackled by a series of detailed case studies and longitudinal investigations.

5 SUMMARY AND CONCLUSIONS

This chapter has reviewed the issues that concern a manufacturing organization as it considers developing a strategy of Mass Customization. We have drawn attention to the need to design the products, inventory strategy, ICT system, performance measures and the supply chain with the new, demanding requirements of Mass Customization in mind. We have discussed the competences that an organization is likely to need in order to be able to carry out effectively a mass-customizing manufacturing strategy. We recommend in Section 4 that the organization must also change in its attitude to its customers and suppliers, since a co-operative, effective supply chain will be needed to meet these new demands.

This chapter has looked at some of the components of a Mass Customization strategy. These include the **manufacturing** strategy, ranging from Fabricate-to-Order through the spectrum to Post-Distribution. The **inventory** strategy must also be considered, since this can affect the levels of customer satisfaction that can be achieved within a near-mass-production lead-time. The **supply chain** strategy includes how to manage the relationship with suppliers under the demanding conditions of Mass Customization, but also how much of the customizing work can be out-sourced to one's suppliers, and the often overlooked issues of common performance measures and schedules. The **business** case must also be developed, in identifying the needs of the customers and adjusting the whole product to meet those needs in a cost-effective manner. We caution that just because another, possibly competing, manufacturer in the same business sector succeeds with a particular overall MC strategy does not imply in any way that it is the only strategy which work in that sector. Companies need to look carefully at their own competencies and goals in order to decide the best way forward.

New research is needed to model mathematically how the Mass Customization process should proceed, but this needs to be carried out simultaneously with longitudinal case studies of organizations as they begin to move from high levels of product variety to mass customization. The best way to do this is likely to be in consortia of manufacturing organizations working together, to share the risk and disseminate the findings as quickly and effectively as possible.

If we look further ahead, i.e. beyond the next decade, we need to bring into the picture, the future form of manufacturing processes and the environmental impact of manufacturing.

One possibility for the future could be the possibility of micro-factories, where production facilities are located in stores or regional centres, much closer to the point of consumption than is currently the case (Wells and Nieuwenhuis 1999). Economies of scale have driven manufacturers to optimize production in mega-factories, which have been designed and built to serve whole continents. The micro-factory may use a modified version of current manufacturing technology to serve a smaller geographical location. This new structure could provide shorter lead-times and a good degree of customization, while possibly reducing the amount of distance that the products and work in progress must undertake.

Another related issue that needs consideration is the environmental costs of product, supported by the reverse logistics chain or the green supply chain. As we seek to drive waste out of the manufacturing process, we may be obliged to consider the whole lifecycle costs of the product, including its disposal. This will require that the constituent parts of the product should be capable of being recycled, re-used or re-furbished and upgraded into another version of the same or a different product. In order to achieve these goals, it is likely that each product will need some sort of life-cycle tagging, to indicate the date, functional specifications and constituent parts of the product. This could create a huge data burden, that will have to be served by the Internet, since the products may be moved anywhere on the planet while still retaining the facility to be re-cycled knowing its original specification. Such requirements will also affect the design of products, their quality and mass customization feasibility.

We have yet to appreciate fully the ways in which the Internet will change manufacturing systems and their supply chains. New approaches to modelling these supply highly inter-connected networks, such as agent-based simulations, will offer new insights on the behavior of manufacturing systems and suggest new strategies for the design and operation of such systems.

There is no doubt that Mass Customization offers a radical and far-reaching new vision of manufacturing, integrating all aspects of the manufacturing system in order to achieve success. We look forward to the future of this new paradigm and await the findings of the research with excitement.

ACKNOWLEDGEMENTS

We acknowledge EPSRC grant number GR/N11926. We also acknowledge discussions and input from Aruna de Alwis, Philip Brabazon, Joanna Braham, Bart McCarthy, Claudia Mchunu, Nuri Shehabuddeen, Ernesto del Valle Lehne and Lu Wuyi.

REFERENCES

1   Ahlström P., Westbrook R. Implications of mass customization for operations management. International Journal of Operations and Production Management 1999; 19: 262-274.

2   Alford D., Scakett P., Nelder G. Mass customization – an automotive perspective. International Journal of Production Economics 2000; 65: 99-110.

3   Duray R. Configurations of Mass Customizers – Descriptions of Manufacturing Infrastructure. Doctoral thesis, Ohio State University, 1997.

4   Duray R., Ward P.T., Milligan G.W., Berry, W.L. Approaches to Mass Customization: Configurations and Empirical validation. Journal of Operations Management 2000; 18: 605-625.

5   Frizelle G.D.M., Suhov, Y.M. An entropic measurement of queuing behaviour in a class of manufacturing operations. Proceedings of the Royal Society; 2001; 457: 1579-1601.

6   Gilmore J.H., Pine B.J.I. The four faces of mass customization. Harvard Business Review 1997; 75/1: 91-101.

7   Holweg M., Pil F.K. Start with the customer. MIT Sloan Management Review 2001; Fall: 74-83.

8   Kotha S. Mass customization: Implementing the emerging paradigm for competitive advantage. Strategic Management Journal 1995; 16: 21-42.

9   Lampel J., Mintzberg H. Customizing customization. Sloan Management Review 1996; 38/1: 21-30.

10   Mchunu C., de Alwis A., Efstathiou J. Decision support framework for establishing a "best fit" mass customization strategy. Working paper, Department of Engineering Science, University of Oxford; 2002.

11   Piller F., Moeslein K. From economies of cale towards economies of customer integration. Technical University of Munich, Department of General and Industrial Management, ISSN 0942-5098; 2002.

12   Pine B.J., Mass Customization: The new frontier in business competition. Boston: Harvard Business School Press, 1993

13   Sako M., Murray F. Modules in Design, Production and Use: Implications for the Global automotive industry. In International Motor vehicle Program (IMVP) Annual Sponsors' Meeting; 1999; Cambridge Mass.

14   Spring M., Dalrymple J. Product customization and manufacturing strategy. International Journal of Operation and Production Management 2000; 20: 441-467.

15   Tseng M., Jiao J. Design for mass customization. CIRP-Annals 1996; 45: 153-156.

16   Victor B., Boynton A.C. Invented Here. , Boston: Harvard Business School Press, 1998.

17   Wells D.P., Nieuwenhuis D.P. Micro factory retailing: a radical business concept for the automotive industry. Automotive Marketing Review 1999; 9/3: 3-9.

18   Zhang T. Modeling and Measuring the Complexity of Mass Customization Systems. Internal Report, Department of Engineering Science, University of Oxford, 2003.

INDEX